OPERATOR'S GUIDE SERIES

Hydraulics for Operators

Barbara A. Hauser

CRC Press
Taylor & Francis Group
Boca Raton London New York

CRC Press is an imprint of the
Taylor & Francis Group, an **informa** business
A TAYLOR & FRANCIS BOOK

CRC Press
Taylor & Francis Group
6000 Broken Sound Parkway NW, Suite 300
Boca Raton, FL 33487-2742

ISBN 13: 978-0-87371-846-2 (pbk)

Library of Congress catalog number: 92-26836

Library of Congress Cataloging-in-Publication Data
Catalog record is available from the Library of Congress

Visit the Taylor & Francis Web site at
http://www.taylorandfrancis.com

and the CRC Press Web site at
http://www.crcpress.com

series preface

This series of operator's guides was conceived by Lewis Publishers and myself as an accessible training tool for water and wastewater plant operators and managers. The guides are not written as detailed, reference-filled treatment studies, but concentrate instead on how to do the actual work required to keep treatment plants running smoothly. The emphasis throughout the series is on proven, common-sense techniques and insight that can help you reliably operate and maintain plant equipment.

The series authors are all experienced professionals who have "been in your shoes." They have walked their share of shifts, know what to watch for, and know how to respond to problems. Their goal is to share this knowledge to ensure efficient operations and improve compliance with current water quality regulations.

Each of the operator guides focuses on a specific unit process or type of work common to water or wastewater plants. The target audience is the men and women who keep it flowing. Every guide is written in straightforward, everyday language and contains information on the following:

- process overview
- maintenance considerations
- component/equipment description
- startup and shutdown procedures
- process control considerations
- typical design criteria and performance

- treatment objectives
- routine operator duties
- troubleshooting guides
- normal operating procedures
- monitoring requirements
- safety concerns

In addition, each contains figures, tables, and a bibliography of suggested references for further study. Any work described in the guides is explained in an easy-to-follow, step-by-step format.

We hope that the continuing evolution and growth of this series will result in more confident and competent operators worthy of increased respect and appreciation as professionals on the front lines of environmental and public health protection.

Glenn M. Tillman
Series Editor-in-Chief

preface

Potable water distribution systems are taken for granted, but they must be carefully designed and controlled by the water utility to provide adequate pressure, safety from backflow contamination, and proper water velocity for protection of household piping.

Water is a carrier for most household and industrial wastes. Proper hydraulic design and maintenance of wastewater collection systems assures sanitary transport to the treatment plant and prevention of sewer backup.

A new operator at a water or wastewater utility is most concerned with learning the daily routine of laboratory testing, operation and process maintenance. The hydraulics of his treatment process has been designed into it. This is what keeps the water moving from one process unit to another, providing correct detention times, proper settling velocity, lift to a higher elevation. Consideration of hydraulic principles may at first be minimal - until the electrical power goes out, a pump malfunctions, a line breaks, or a blockage occurs. When hydraulic control is no longer there, treatment process grinds to a halt. No water utility can operate without it.

In as simple a manner as possible, this text covers the principles and calculations dealing with the hydraulics of water systems. It stresses what is necessary for a basic understanding, and emphasizes practical applications for water and wastewater utility operations.

The principles are explained within the chapters, and all practice problems are placed at the end of the text.

The reader should be aware that:

- A scientific calculator will be needed.
- Some previous mathematical knowledge will be assumed, as will knowledge of a few formulas: area of a circle and rectangle, volume of containers of these shapes.
- Most values are converted to units of "feet" for calculation. (This includes: feet, square feet, feet per second, cubic feet per second.)
- Most answers are rounded off to one or two decimal places.
- In diagrams, the pressure gage symbol with a number inside it designates a pressure reading in psi (pounds per square inch).
- Practice problems listed at the end of the text cover a range of difficulty, in order to adapt to different level state certification examinations.

author

Barbara Hauser has been employed as department head and chief instructor of the Water Purification Technology department at Bay de Noc Community College in Escanaba, Michigan for the past nine years. Her background includes biochemistry research with the Rockefeller Institute in New York City and at the University of Maine in Orono. She has worked for the Rexnord Company, Milwaukee, Wisconsin as an industrial operations specialist in advanced wastewater treatment, and for the city of Gladstone, Michigan in municipal wastewater treatment. She is a licensed operator in water and wastewater treatment in the states of Michigan and Wisconsin.

In addition to her academic duties, Mrs. Hauser has developed and operated onsite training courses for industrial wastewater treatment personnel, instructed short courses and correspondence courses for waterworks operators sponsored by the Michigan Department of Public Health, organized regional waterworks and wastewater operators training meetings, conducted research for the Michigan Department of Natural Resources, published "Practical Hydraulics Handbook", the predecessor of this manual, and has raised four children.

Mrs. Hauser has an Associate of Applied Science degree from Bay de Noc Community College in Water Technology, a Baccalaureate Degree in Biology and Chemistry from Trinity College in Washington DC, and a Master of Science in Occupational Education from Ferris State University, Big Rapids, Michigan. In 1990, she was awarded the "Outstanding Occupational Educator" award by the Michigan Occupational Dean's Council.

acknowledgments

I am grateful to the faculty and staff of Bay de Noc Community College. Their encouragement has carried me far beyond my expectations.

note

contents

1 MASS, DENSITY AND FLOW 1
Mass & Weight 1
Density 1
Temperature 2
Pressure 2
Specific Gravity 2
Displacement 3
Flow 4
Equation of Continuity 5

2 PRESSURE 7
Pascal's Law 9
Container Shape 10
Pumps 11
Pressure Measurement 12
  Piezometer 12
  Pressure Gage 12
Gage Position 13
Air Pressure/Water Pressure 14
  Atmospheric Pressure 14
  Absolute Pressure 14
Total Pressure - Force 15
Effects of Pressure - Dynamic Systems 16
  Water Hammer 16

Thrust 17
Thrust Blocks 18
Thrust Anchors 18

3 BERNOULLI'S THEOREM 19
Static Systems 19
Dynamic Systems - Head Loss 21
Types of Head 21
Pressure Head 21
Elevation Head 22
Velocity Head 23
Total Head 26
Head Loss 26
Steady State Flow 26
Bernoulli's Theorem 27

4 PUMPING - INTRODUCTION 33
Pumping Heads 33
Static Heads 34
Dynamic Heads 35
Power Calculations 37
Well Problems - Pumping Problems 40
Groundwater Well 40
Well Yield 40
Static Water Level 40
Pumping Water Level 40
Drawdown 40
Safe Well Yield 40
Radius of Influence 41
Cone of Depression 41
Specific Capacity 41
Wet Wells 41
Water Level Sensing and Pump Control 42
Mechanical Floats 42
Diaphragm Element 42

Bubbler 42
Electronic Probes 42
Ultrasonic Detectors 42

5 FRICTION LOSS 43
Components of Head Loss 43
Roughness 43
Length 44
Diameter 44
Velocity 44
C Factor 44
Slope 46
Compound Pipes 49
Calculating with Pipes in Series 49
Calculating with Pipes in Parallel 49
Minor Head Loss 50
Equivalent Pipe Technique 51
Velocity Head Technique 53
Ordinary Exit 54
Ordinary Entrance 54
Borda Entrance 54

6 OPEN CHANNEL FLOW 55
Calculations for Open Channel Flow 57

7 FLOW MEASUREMENT 63
Flow Rate 64
Flow Amount 64
Flow Rate Meters 64
Pressure Differential Meters: Closed Pipe Flow 64
Orifice Meters 64
Venturi Meters 66
Flow Rate Controller 67
Differential Manometers 68

Pressure Differential Meters: Open Channel Flow 70
Weirs 70
Rectangular Weir 70
90 Degree V Notch Weir 70
Parshall Flumes 70
Inlet 72
Throat 72
Outlet 72
Trajectory of a Jet 72
Pitot Gage (Pitot Tube) 73
Magnetic Flow Meter 74
Ultra Sonic Meters 74
Rotameter 75
Totalizer Meters 75
Positive Displacement Meters 75
Turbine Meters 76
Compound Meters 77
Propeller Meters 78
Proportional Meters 78
Sizing Service Lines and Meters 78

8 CENTRIFUGAL PUMPS 87
Pump Types - Position 89
Vertical Pump 89
Horizontal Pump 90
Pump Types - Exit Flow Direction 90
Radial Pump 90
Axial Pump 90
Pump Types - Suction Condition 90
Single Suction 90
Double Suction 90
Pump Types - Impellers 90
Single Stage 90
Multi-Stage 91

Other Important Pump Components 91
Stuffing Box and Packing 91
Mechanical Seals 91
Rotating Assembly 91
Stationary Assembly 91
Bearings 92
Radial Bearing (Line Bearing) 92
Thrust Bearing 92
Pumping Characteristics: The Pump Performance Curve 92
Changing System Conditions 95
The System Curve 96
Pumps in Series 97
Booster Pumping 97
Multi-Stage Pumps 97
Pumps in Parallel 98
Changing Pump Performance 98
Change Pumping RPM 99
Change Impeller Size 99
Variable Speed Pumps 99
Priming a Pump 100
Foot Valve 100
Vacuum Pumps and Ejectors 100
Priming Chamber 101
Cavitation 101
Atmospheric Pressure 103
Velocity Head 103
Suction Lift 103
Suction Losses 103
Vapor Pressure 103
Entrained Air 104
Insufficient Submersion 104
Shallow Suction 104
Slope of Associated Piping 104
Reducers 104

Specialized Pumping Systems 105
Submersible Pumps 105
Deep Well Pumps 107
Jet Pumps 107
Pump Appurtenances: The Pressure Gage 107
Pump Operation Guidelines 108
Selection 108
Installation 109
Operation 109
Maintenance 109

EXERCISES 111

GLOSSARY 135

BIBLIOGRAPHY 143

INDEX 145

OPERATOR'S GUIDE SERIES

Hydraulics for Operators

1

mass, density and flow

mass & weight

Mass is the quantity of matter that a substance contains. It is a basic property of the substance, and is constant, regardless of location. Mass is most frequently registered in units of grams, or pounds; it is measured with an analytical balance which compares it against a known mass.

Weight is the effect of the force of gravity upon a substance, and is measured with a scale. A block of steel which weighs ten pounds on earth, will weigh much less on the moon, where gravitational forces are less. Its mass, however, will be the same in both places.

Since we are dealing only with earthly water systems, for our purposes mass and weight are the same quantity.

density

Also called SPECIFIC WEIGHT, density is mass per unit volume, and may be registered as lb./cu.ft., lb./gal., grams/cu.meter. Taking a fixed volume container, filling it with a fluid, and weighing it, we can determine density of the fluid (after subtracting the weight of the container).

1 cubic foot of water weighs 62.4 pounds.
(Density=62.4 lb./cu.ft.)

1 gallon of water weighs 8.34 pounds.
(Density = 8.34 lb./gal.)

1 milliliter of water weighs 1 gram.
(Density = 1 gram/ml.)

temperature

The density of materials is affected by temperature. Many solids expand when hot. Fluids also become less dense when warm. Water has the peculiar property of being most dense at four degrees Celsius. Above and below that temperature, it expands. This accounts for seasonal turnover in lakes, and the formation of density gradients, or layers, in reservoirs and behind dams, often concentrating pollution at specific depths. In settling tanks and clarifiers, this effect can inhibit uniform mixing of influent water, causing short circuiting. Most solids will settle more effectively in warm waters than in cold waters because the water is denser when cold.

pressure

The density of most liquids is not affected by pressure. For all practical purposes, water is considered an incompressible fluid; 62.4 pounds of water occupies a volume of one cubic foot, regardless of the pressure applied to it.

specific gravity

This value designates the ratio of the density of a substance, compared to the density of a standard substance. For liquids and solids, the standard chosen is the weight of water. Therefore:

$$\text{Specific Gravity} = \frac{\text{Density of Substance}}{\text{Density of Water}}$$

If the Specific Gravity of water were to be determined this way, we would write:

$$\text{Specific Gravity} = \frac{\text{Density of Water}}{\text{Density of Water}} = 1$$

Substances with a Specific Gravity of less than 1 will float (gasoline, styrofoam, wood, wastewater scum). Substances with a Specific Gravity of more than 1 will sink (bricks, steel, grit, floc, sludge). A good example of treatment process taking advantage of different Specific Gravities is the multi-media filter. During backwash, the different grades of media are mixed up, but at the end of the backwash cycle, as the water quiets, the media layers become separate and distinct: anthracite on top, sand in the middle, garnet on the bottom. These substances are chosen for filtration partially because of their different Specific Gravities.

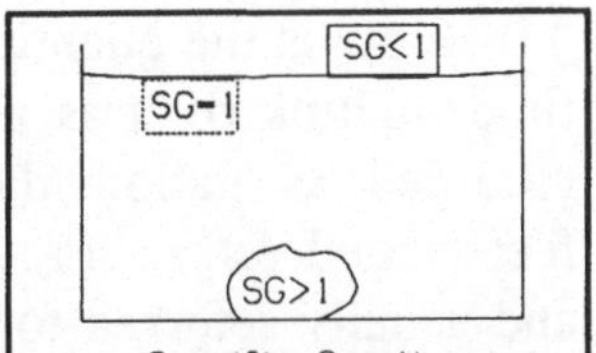

Specific Gravity

An anaerobic digester is another good example of the effect of Specific Gravity. Separation of solids, liquid and gas is essential to operation, so that each can be drawn off separately.

displacement

A solid submerged in a body of water will displace its own volume. The displacement causes a rise in water level equivalent to the volume of the submerged object. An object partially submerged displaces a volume of water equivalent to the volume of the section.

Sometimes it is useful to know the volume of an irregularly shaped object. This can be determined by measuring the displacement. For example, to calculate the percent volume of mudballs in a sand filter, take a core sample with a small cylinder of known volume. Extracting a portion of media, sift out the sand, leaving only the mudballs. Pour them into a large graduated cylinder, filled to a liter with water. A volume of water will be displaced in the cylinder equivalent to the volume of the mudballs. Record the milliliter rise in the cylinder, divide by the volume of your core sampler, and you have your percent mudballs - in your sampler, and in your sand filter.

flow

Flow is the quantity of water passing a point in a given unit of time. Think of it as a volume - which is moving. Flow can be recorded as gallons/day (gpd), gallons/minute (gpm), or cubic feet/second (cfs). Flows treated by a municipal utility are large, and usually referred to in Million Gallons per Day (MGD).

Flow may be enclosed in pipes, or it may be open channel flow. The first five chapters of this text will be considering predominantly closed pipe flow. Pipes are flowing full, and under pressure. Open channel flow deals with conduits which are partially full, where the water surface is in contact with atmospheric pressure (streams, aqueducts, sewer pipes, grit chambers, clarifiers). The water flows because of the slope of the conduit. Open channel flow will be taken up later in the text.

Flow may be laminar or turbulent. Laminar flow only occurs at extremely low velocities. The water moves in straight parallel lines, called laminae, or streamlines, which slide upon each other as they travel, rather than mixing up. Turbulent flow, which is normal pipe flow, occurs because of friction encountered on the inside of the pipe. This throws the outside layers of water into the inner layers; the result is that all the layers mix and are moving in different directions, and at different velocities. Added up, however, the direction of flow is forward.

Flow may be steady or unsteady. We will be considering steady state flow only. At any one point, the flow, and velocity, does not change. Most hydraulic calculations are done under the assumption of steady state flow. If we can assume a given flow at a given point, calculations can be based on that fact. If we had to consider that the flow was constantly changing, calculations would become extremely complex. Note, however, that in actual systems, flow is frequently in an unsteady condition. As a water tank empties into a pipe, the flow through the pipe decreases over time, because the depth of water in the tank which pushes the water forward is decreasing. As a pump fills a tank from a bottom entrance, that pump is delivering less water as the tank gets fuller and fuller. Flow through a wastewater treatment plant is constantly fluctuating, but process treatment units are designed for an average, maximum or minimum flow, depending on the need. We realize that unsteady flow is a common condition, but we do not bring this into calculations.

equation of continuity

Under the assumption of steady state flow, the flow that enters the pipe is the same flow that exits the pipe. It is continuous. Water is incompressible; it cannot accumulate inside. The flow at any given point is the same flow at any other given point on the pipeline. This is true in any water system, as long as no additional flows are added, and no exits which split or divert the flow away in other directions are present.

However, the velocity of the water may change. At a given flow, the velocity is dependent upon the cross sectional area of the conduit. Velocity is the speed at which the flow is traveling. A given quantity of water will travel faster through smaller spaces, and slower through larger spaces. Consider this phenomenon in a closed pipe. A 50 gal./min. flow travels at a very slow velocity through a 36 inch diameter pipe, but if the pipe narrows to 4 inches diameter, that same 50 gal./min. will be moving very fast through the narrow section, in order for the flow to remain the same.

The principle applies equally to open channel flow. Water entering a wastewater treatment plant travels at scouring velocity coming down the pipe (2-3 ft./sec.). Entering the grit chamber, which is larger in cross sectional area than the sewer pipe, the velocity slows down to about 1 ft./sec. so that organics will settle. Much of wastewater treatment is designed upon this principle, and detention times are developed from it.

Hence was developed the <u>Equation of Continuity</u>:

$$Q = A\ V$$

Where:

Q = Flow
A = Area (cross sectional area of conduit)
V = Velocity

This is the most basic hydraulic equation. It always holds true, and all other hydraulic formulas determine components of this one.

Calculations are very straightforward, solving for the unknown factor. Be aware that if pipes of different diameters are considered, substitution within the formula can be performed.

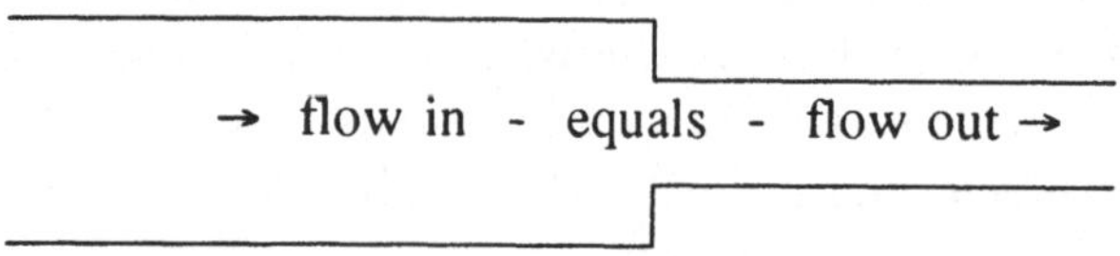

$$\uparrow \qquad\qquad \uparrow$$
$$Q_1 = Q_2$$

Therefore:

$$A_1\ V_1 = A_2\ V_2$$

A calculation for velocity, or for pipe diameter, can often be done more quickly using this version of the formula.

> **Note:** The unit labels must be in the same dimensions across the formula. It is easiest to label flow as cu.ft./sec. (cfs), area as sq.ft., and velocity as ft./sec. If values are presented in other dimensions, first change them to these before inserting into the formula.

2

pressure

As all fluids, water has a specific weight: 62.4 pounds for every cubic foot, or 8.34 pounds for every gallon. This weight, resting on a surface, exerts a force on that surface. For example, one cubic foot of water, resting on its bottom surface, exerts a force of 62.4 pounds on that square foot (62.4 lb./sq.ft.). Force on a unit area is called <u>Pressure</u>.

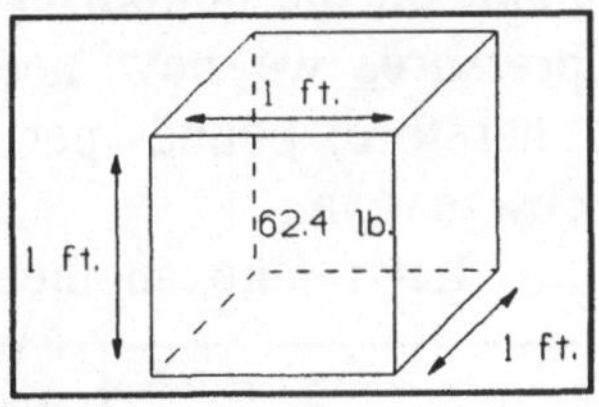

> **Note:** The unit label for specific weight (density) is a weight per volume designation (lb./cu.ft.). The unit label for Pressure is a weight per area designation (lb./sq.ft.). It applies to the surface area that the weight is resting on.

If this weight were composed of two cubic foot blocks of water, one on top of the other, then the specific weight would be 62.4 lb./cu.ft., the height would be 2 feet, and the pressure on the bottom surface would be 124.8 lb./sq.ft. Density remains constant, but the pressure on the unit area depends on the height, or depth of the water.

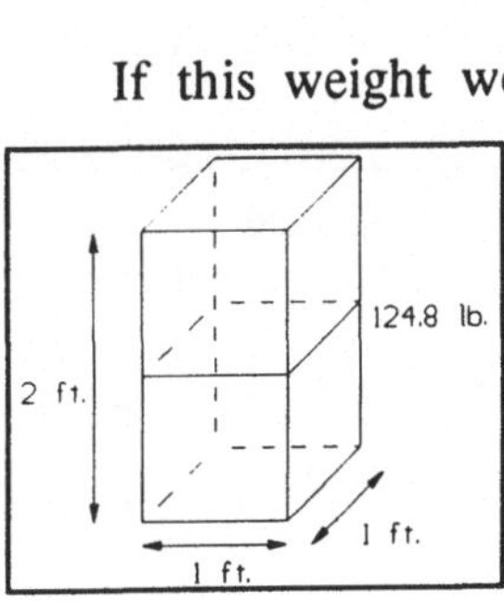

A formula develops:

Pressure = Weight x Height

124.8 lb/sq.ft. = 62.4 lb/cu.ft. x 2 ft.

If there were four blocks of water, arranged as in the diagram below, the pressure on every square foot of bottom surface would still be 124.8 lb./sq.ft. It would not matter how wide the water was. It is just the height, or depth, that is pushing down on that surface, and causing the pressure.

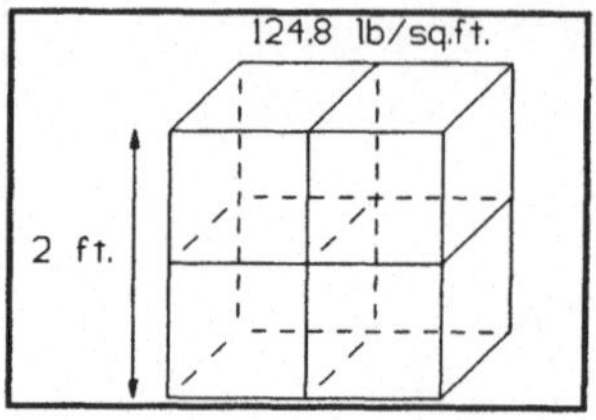

It is most convenient to specify pressure at a particular point in a water system. For example, what is the water pressure at a specified point along a pipeline? Or, what is the pressure at a particular point on the bottom of a swimming pool? Using the concept of pressure, we now need to deal with a very small unit area. Therefore, pounds per square inch (psi) has become the favored designation.

Let's take another look at the cubic foot block of water.

Wishing to register pressure as lb./sq.in., the bottom surface must be split into 144 square inches. Now, how much water is resting on each of those square inches? Picture a column of water rising from each of them to the top of the block. How much does each column weigh?

$$\frac{62.4 \text{ lb./cu.ft.}}{144 \text{ sq.in./sq.ft.}} = .433 \text{ lb. per foot of height}$$

<u>A cubic foot block of water, one foot high, exerts a pressure of .433 lb. on every square inch of bottom surface area</u>.

The double block of water, two feet high, exerts a pressure of .866 lb. on every square inch of bottom surface area. In all cases, the pressure depends upon the depth of the water above it, and an equivalence of feet of depth to psi pressure is created.

However, there is a more comfortable way to state this equivalence. For practical purposes, the water utility superintendent would like to know how many feet of water depth he has to provide in the city water storage tank in order to have adequate pressure in the pipes. There is needed an equivalence of feet of water depth to psi pressure that is readily interchangeable, and which will yield discrete units of psi pressure.

Creating a ratio, if a 1 ft. column of water is equivalent to a pressure of .433 psi, then how many feet of water will equal 1 psi?

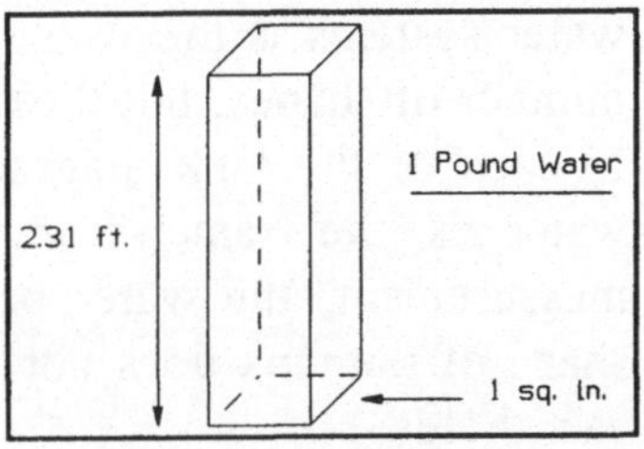

$$\frac{.433 \text{ psi}}{1 \text{ ft.}} = \frac{1 \text{ psi}}{x \text{ ft.}}$$

$$x = 2.31 \text{ ft.}$$

Therefore, we see that <u>1 psi is equivalent to a 2.31 ft. depth of water</u>. Water pressure can be registered as psi, or as feet of water. The conversion 1 psi = 2.31 ft. is most commonly used. However, .433 psi = 1 ft. will give the same result in calculation.

pascal's law

In the 17th century, Blaise Pascal formulated a fundamental law of hydraulics:

<u>The pressure at any one point in a static liquid is exerted with equal intensity in all directions.</u>

For instance, the pressure at any point on the horizontal bottom of a tank of water is the same, and the pressure exerted upward, downward, and to the sides from any one point is the same also. Dewatered basement floors in flooded areas buckle from this upward pressure, and structures built underground, such as dewatered tanks at water or wastewater treatment plants, must be able to withstand upward pressures imposed by groundwater, or they may float right up out of the ground.

container shape

Using both of these facts:

- Pressure in a static system is due to the depth of water above the point of measurement.
- Pressure is exerted with equal intensity in all directions.

The source of distribution system pressure in most community water systems is the water storage tank. These tanks may take a number of shapes, but these same basic hydraulic principles apply. In each of the tank diagrams below, as long as the depth of the water is the same from the water surface to the point of measurement, the water pressure at that point is the same. The shape of the tank does not matter, nor does the number of gallons each holds.

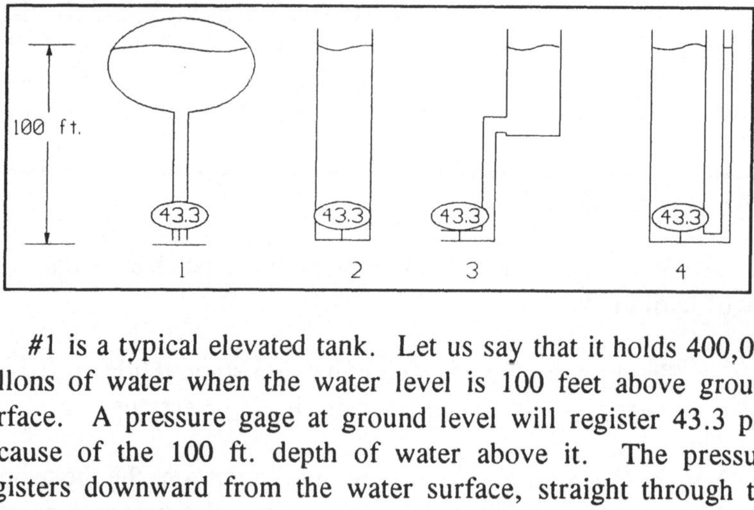

#1 is a typical elevated tank. Let us say that it holds 400,000 gallons of water when the water level is 100 feet above ground surface. A pressure gage at ground level will register 43.3 psi, because of the 100 ft. depth of water above it. The pressure registers downward from the water surface, straight through the riser pipe to the gage.

#2 is a standpipe, also containing 400,000 gallons of water when the surface is 100 ft. above ground. It will also register 43.3 psi at the bottom. The difference between these two tanks is in the use of the water. The water pressure which supplies the community from #1 will remain high until the tank is empty. #2 provides the same pressure when full, but as water is used and the level drops, the pressure will drop significantly with it.

#3 is a ground level storage tank, set into a hill, with a pressure gage on the pipeline 100 ft. below the water surface. The pressure gage will register 43.3 psi. The community is down at this level, and the water storage tank was installed at a higher elevation to provide the pressure. It doesn't matter that the tank is buried in the ground instead of being up in the air. The same depth of water is still provided by the direct physical connection of water from the tank through the pipeline. The pressure registers down, and to the side, all the way down the pipe to the gage location.

#4 is a copy of #2, equipped with a 1/2 inch diameter sight glass, to read the water depth. Connected to the bottom of the tank, the water will rise in the glass to the water level in the tank - 100 ft. The gage at the bottom of the tank registers 43.3 psi. If a valve between the sight glass and the tank is then closed, a gage at the bottom of the sight glass will still register 43.3 psi. Though there is only a little water in that glass, it is 100 ft. deep above the gage.

pumps

Adequate elevation and depth of water in a tank will provide enough pressure to move the water throughout the distribution system.

PUMPS ALSO PROVIDE PRESSURE!

When elevation is not available, a pump is used to provide the needed pressure. The pump is actually providing lift (elevation), or enough pressure to raise the water to a given height. If the direction of flow is horizontal from the pump discharge, then the pressure remains in the discharge pipes, and keeps the water moving. Pumped pressure can also be registered in feet of water, or in psi.

pressure measurement

Piezometer

The simplest type of pressure measurement device is the piezometer. This is simply a narrow, open-ended vertical tube, attached at the point of pressure measurement. Water will rise in the tube to a height equivalent to the pressure. If a piezometer is attached to the bottom of a tank, and is made of glass, it is referred to as a sight glass, and provides an easy means of reading the water level. The water depth can easily be converted to psi to obtain a pressure reading at the bottom. This is practical for small tanks of water at ground level, but obviously impractical on an elevated storage tank. Piezometers can be installed on pipelines. The water will rise in the tube to a level equivalent to the pressure in the pipe. As long as the piezometer is tall enough, the water will not pour out the top. It will rise till the depth of water in the tube provides a downward pressure equivalent to the upward pressure in the pipe. Again, this is most often impractical. A pipeline under a pressure of 60 psi, would need a piezometer 138.6 ft. high.

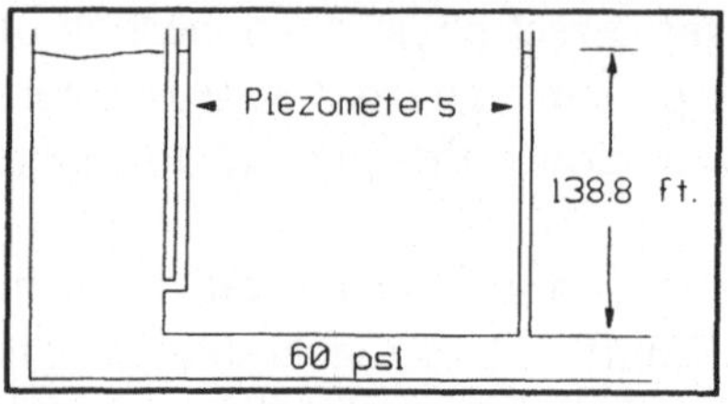

Pressure Gage

The pressure gage is a compact, practical device used for pressure measurement. It is probably the most important instrument in the water system. Called a Bourdon Tube, it should be calibrated in feet of water, though it may read feet, psi, or inches of mercury. The dial size is usually 4 1/2 inches in diameter, for readability, and the range selected should go beyond

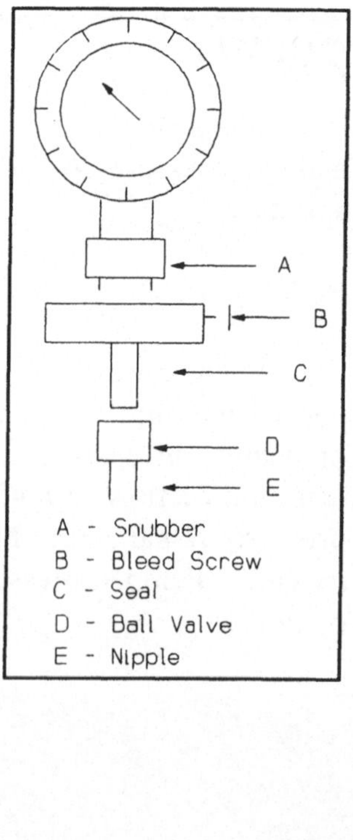

the maximum operating pressure of the system. The needle can be zeroed with a screwdriver. The gage assembly usually is a liquid filled system. The snubber is a restrictive device that stops movement of the liquid fill, deadening pulses of power which cause a blurred pointer motion. Glycerine is often preferred for fill, and prevents solids bearing water from damaging the gage. The diaphragm seal isolates the water from the fill, but allows the pressure to be transmitted through. The bleed screw on the bottom of the diaphragm seal is for bleeding air if the pointer does not return to zero when the gage is disconnected. A ball valve should be at the bottom for shutoff. The nipple is the connection of the gage to the pipe, and should be welded on or attached with a service saddle (not tapped in). When this area becomes clogged, it is easy to remove the gage above the valve, and rod out the nipple and valve.

gage position

A pressure gage is usually attached directly onto the pipe, and reads pressure directly from it. Occasionally, for ease in reading, the gage is placed at some distance from the pipe, either above or below it. If an automatic correction for the difference in elevation is not built into the apparatus, it is easy to calculate. A gage placed below the pipeline will register a higher pressure than the pressure in the pipe. The water extends down to the diaphragm seal on the gage. If the extension is 5 ft. long, then the gage will register the pipe pressure, plus the extra five feet of elevation. If the gage is above the pipeline, it will register the pipe pressure minus the distance between the gage and the pipe. In this case the pipe pressure is more than the gage pressure reading.

> **Note:** Be careful to remember to change pressure units from psi to feet when calculating in these distances.

Keep in mind that even if the water in the pipe is flowing, the water exiting to the pressure gage is static, it dead ends at the gage. This water is not moving, and except for the difference in elevation from pipe to gage, the actual pipe pressure at the point of connection is being transmitted to the gage. Bends and turns in the gage extension line will not affect the gage reading (Pascal's Law - pressure registers equally in all directions). Only the elevation difference between gage and pipeline is significant.

air pressure / water pressure

The pressure read on a pressure gage is referred to as Gage Pressure. It is the difference between a given pressure and that of the atmosphere. It is most often a positive number, but can also be negative, such as in siphon lines, or on the suction side of a pump which draws water up from below. Some gages are calibrated to read negative pressures, and register it with a minus sign. These gages are reading the amount of pressure below atmospheric pressure; they are actually reading the amount of vacuum in the line.

Atmospheric Pressure

The atmosphere is composed of gases. They have weight, as liquids do, and exert a pressure upon the earth, approximately 14.7 psi at sea level. If all the air were vacuumed out of a container, the

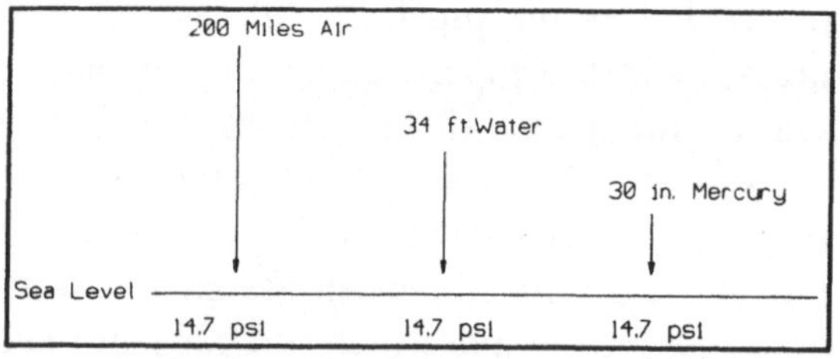

pressure reading on an attached gage would be -14.7 psi, or -34 feet of water: Absolute Vacuum. This is equivalent to an Absolute Pressure reading of 0.

Absolute Pressure

Pressure readings can be reported either as Absolute Pressure or as Gage pressure. The difference is that in Gage Pressure readings, atmospheric pressure is not counted. With Absolute Pressure readings, it is.

Absolute Pressure (psia) = Gage Pressure + Atmospheric Pressure (14.7 psi)

Absolute Pressure is always positive. It is theoretically impossible to have an Absolute Pressure reading below zero.

Gage Pressure is the pressure above or below atmospheric, and is sometimes recorded as psig. It can be negative - down to minus 14.7 psig. The Absolute System of measuring pressures is less commonly used in hydraulic calculations, and this text will deal with Gage Pressures unless otherwise indicated.

total pressure - force

Force is the pressure on an entire area. The force downward on the entire bottom of a tank filled with water, laterally on the side of a dam, or upward on the cover of an anaerobic digester. Force is registered in pounds, and is calculated by multiplying the pressure (lb./sq.ft.) times the area (sq.ft.).

Force = Pressure x Area

lb. = lb./sq.ft x sq.ft.

The formula can be used with other unit labels (square inches) as well, as long as the same dimensional units are carried across the entire formula.

Our original cubic foot of water exerts a <u>pressure</u> of .433 lb. on each square inch of bottom surface. It exerts a <u>force</u> of 62.4 lb. on the entire square foot of bottom surface.

When calculating force on a vertical surface, such as the side of a tank, an average must be taken. The pressure at the water surface is zero; the pressure at the bottom is maximum; therefore, an average of the two must be used, and then multiplied by the sidewall area.

Force becomes significant when dealing with process tanks which are set below ground surface. If the water table is high in the area, it may be necessary to dewater the area around the tank before dewatering the tank for cleanout or repairs, or the surrounding force of water exerted upward on the bottom of the empty tank will pop it right out of the ground.

When a force is applied to a gas, the gas is able to absorb the pressure, and compresses to a smaller volume. When the pressure is released, more space is available, and the gas expands to fill the entire space. For practical purposes, liquids do not have this capacity. They are incompressible.

WATER IS AN INCOMPRESSIBLE FLUID.

When pressure is applied, it will not condense; volume remains the same, and pressure is transmitted to the surroundings. This yields dramatic results when moving water comes to a sudden halt.

effects of pressure - dynamic systems

Water Hammer

Called Hydraulic Shock, water hammer is the momentary increase in pressure which occurs in a moving water system when there is a sudden change of direction or velocity of the water.

When water flowing in a pipeline is suddenly stopped by a rapidly closing valve, pressure energy is transferred to the valve and pipe wall. Shock waves are set up within the system. Waves of pressure travel backwards until encountering the next solid obstacle, then forward, then back again. Neither the pipe nor the water will compress to absorb the shock. Water is 100 times less compressible than steel; this sudden increase in pressure will damage pipes, valves, and shake fittings loose. The velocity of a pressure wave is equal to the speed of sound; therefore it "bangs" as it travels back and forth, until dissipated by friction losses. We are familiar with the "bang" that resounds throughout an older house when a faucet is suddenly closed - an effect of water hammer.

Air chambers are installed in areas where water hammer is encountered frequently, and are typically seen behind sink and tub fixtures, and on primary sludge piston pumps. Shaped like thin upside-down bottles, with a small orifice connection to the pipe, they are air filled. The air compresses to absorb the shock, protecting the fixture and piping. On centrifugal pumps, which can be damaged by hammer caused when electrical power fails, the best form of prevention is to have automatically controlled valves which close slowly.

A less severe form of hammer is called Surge, a slow motion mass oscillation of water caused by pressure fluctuations in the system. This could be pictured as a slower "wave" of pressure building within the system. In long pipelines, surge can be relieved with a surge tank, directly connected to the pipeline. When surge is encountered, it will act to relieve the pressure, and can store excess liquid. Surge tanks can serve for both positive and negative pressure fluctuations.

Thrust

Thrust is the force which water exerts on a pipeline as it turns a corner. The phenomenon is similar to water hammer, except that in a moving system, it is happening constantly. Caused by an imbalance of internal pressures, the intensity of thrust is dependent upon momentum (mass times velocity). The larger the pipeline, the faster the water is moving, the more the thrust. The force can be very great in large pipes, even when pressures are low. Thrust acts perpendicular to the outside corner of the pipe. It affect bends, tees, reducers, dead ends, and tends to push the coupling away from both sections of pipeline. Uncontrolled, thrust will separate the coupling, and cause the pipe to leak or break. Two devices are in use for thrust control in larger pipelines.

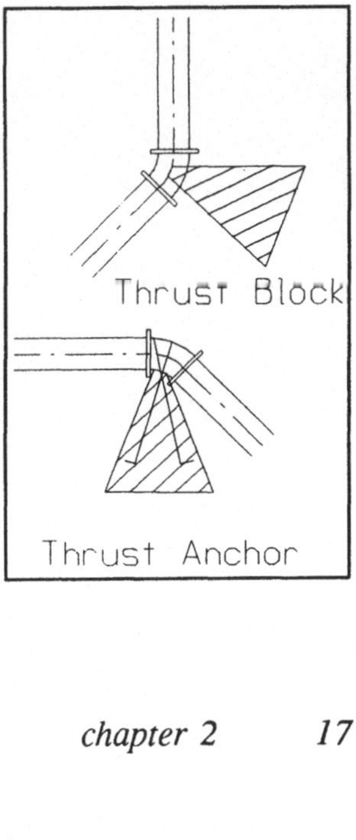

Thrust Blocks - a mass of concrete cast in place onto the pipe and around the outside corner of the turn. These are used for pipes with an elbow or tee which turns right or left or slants upward. The thrust force is transferred to the soil through the larger bearing area of the block.

Thrust Anchors - a restrained joint system employing a massive block of concrete cast in place below a downward bend and shackled to it with steel rods.

The size and shape of a thrust control device depends on pipe size, type of fitting, water pressure and velocity, expected hammer and soil type. In places where these cannot be used, restrained joint pipe is an alternative; it directly transfers the load from the pipe to the surrounding soil.

3

bernoulli's theorem

static systems

We have seen that the water surface in an open vertical tube installed onto a tank of water will rest at the same level as the water surface in the tank. The tube is called a PIEZOMETER. The water level to which the water rises in the piezometer is called the PIEZOMETRIC SURFACE.

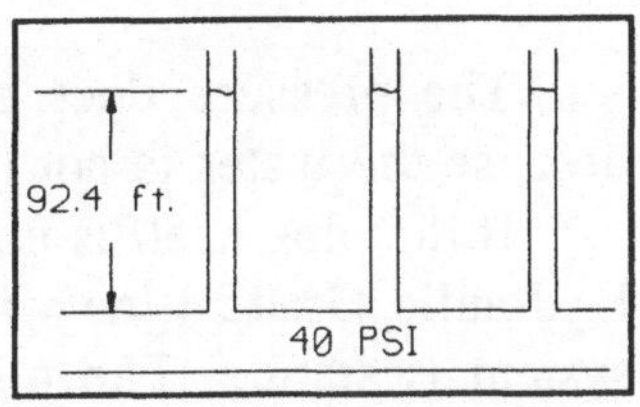

Now consider a pipeline under a constant source of pressure, with a closed valve at the end (static system). If several piezometers are installed onto the pipe in series, the water will rise in each of them to a height which is equivalent to the pressure in the pipe, and it will be at the same level in each of them, because the system is static. If a line were drawn from one piezometric surface to another, all along the pipeline, that line would register the level of pressure energy over the length of that pipe. It is called the HYDRAULIC GRADE LINE.

THE HYDRAULIC GRADE LINE IS THE LINE THAT CONNECTS THE PIEZOMETRIC SURFACES ALONG A PIPELINE.

In the diagram below, the pressure source is included, with its own piezometer attached. The pressure in the pipeline has originated directly from the depth of water in this tank, and each of the piezometers is registering this pressure. Note the closed valve on the pipe end, designating a static water system.

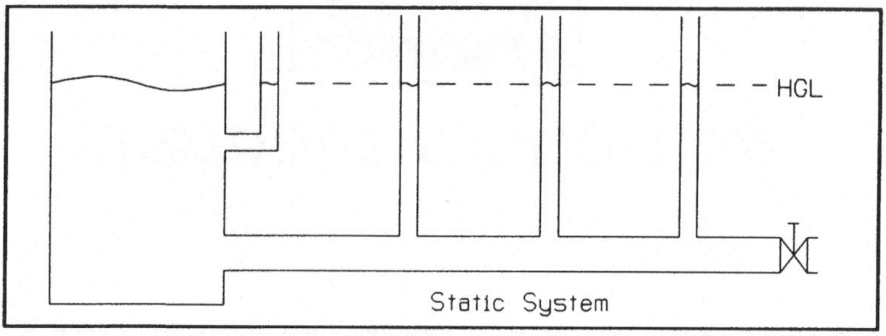

IN A STATIC WATER SYSTEM, THE HYDRAULIC GRADE LINE IS ALWAYS HORIZONTAL.

The pressure does not change along the length of the pipe because the water is not moving and there are no friction losses.

If the pipe itself is not horizontal, but curves up and down, the Hydraulic Grade Line will still remain horizontal, as long as the system is static. The pipeline pressure will be lower where the pipe is elevated, and higher where the pipe is at a lower level.

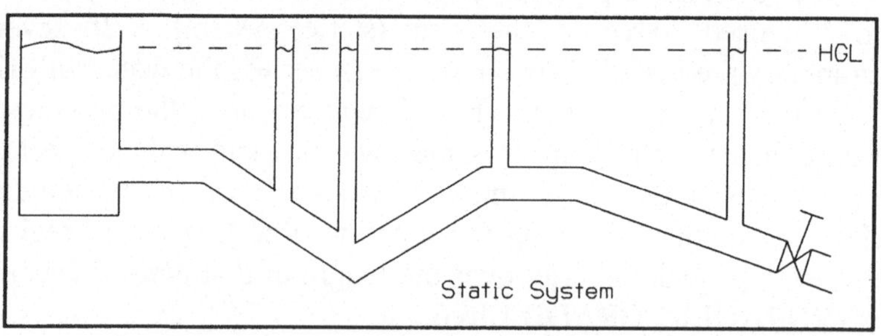

The concept of a piezometric surface is not only demonstrated in man-made pipe systems, but also in nature, with groundwater. A well, which is actually a piezometer, is drilled into the aquifer. If this is a water table aquifer, and not under pressure, water will enter the well to the level of the water table. If the well has been drilled into an artesian aquifer, under pressure, then the water will rise in the well to a distance equivalent to the pressure in the aquifer at that point, possibly all the way to the surface.

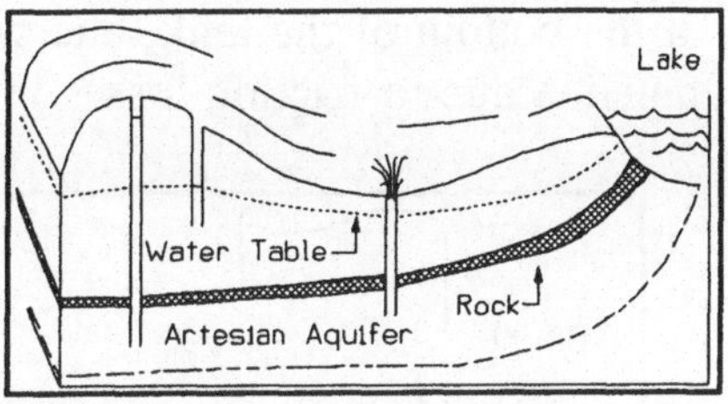

dynamic systems - head loss

In a Dynamic Water System, where water is flowing, the piezometric surface varies. The Hydraulic Grade Line slopes downward along the length of pipe, showing a loss of pressure energy. The difference between the level of the Hydraulic Grade Line in the static system, and that in the dynamic system at any one point, is the feet of pressure lost because of friction in the pipe. This is referred to as HEAD LOSS, and is an important factor. A water utility must account for these pressure losses in order to provide adequate pressure service throughout the system, when determining the water level in the municipal water tower.

types of head

Pressure at a given point originates from the height, or depth of water above it. It is this pressure, or head, which gives the water energy, and causes it to flow. The components of pressure are:

Pressure Head

This is the pressure which is directly due to the depth of the water. If a tank of water is 80 ft. deep, the head is 80 ft. at the bottom of the tank and the psi pressure is 80 ft./2.31 ft/psi = 34.6 psi. The same pressure registers at all points along a horizontal connected pipeline, as long as the system is static. If it is a dynamic system and water is flowing, the pressure will be 34.6 psi at the bottom of the tank, and will decrease over the length of the pipeline due to friction losses in the pipe.

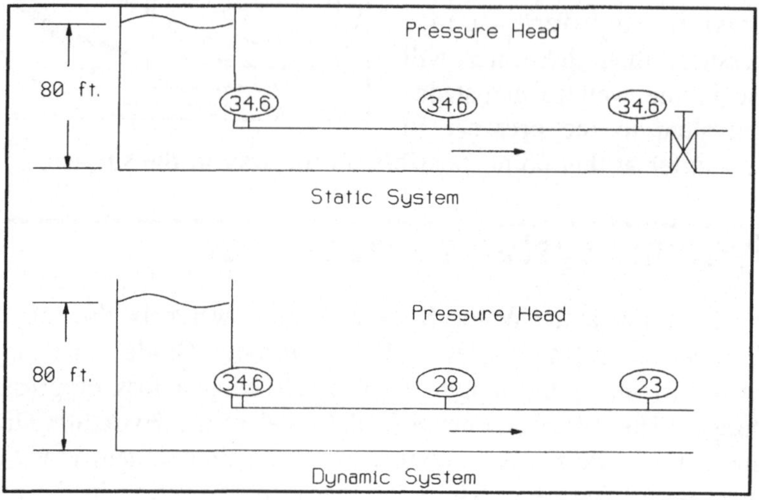

> **Note:** Pressure Head, when used in calculations, is always registered as feet of water.

Elevation Head

If the tank of water (80 ft. deep) were up on a hill, higher in elevation than the point of measurement, the pressure in the static system would still be measured from the water surface (HGL is

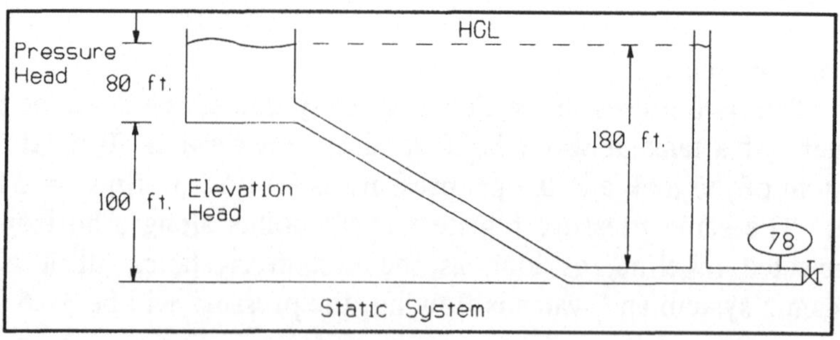

always horizontal in a static system). From the point of measurement at the lower elevation, a pressure gage reading would include the depth of the water, and the elevation of the tank. This pressure due to elevation registers on the gage and is called Elevation Head.

Water storage tanks are often placed at high elevations in order to take advantage of Elevation Head. Some large city water utilities have chosen to draw their raw water from more distant locations, rather than a closer source, because of the available Elevation Head to provide added pressure. Most cities have elevated water storage towers, which also employ Elevation Head to provide adequate pressure to the community.

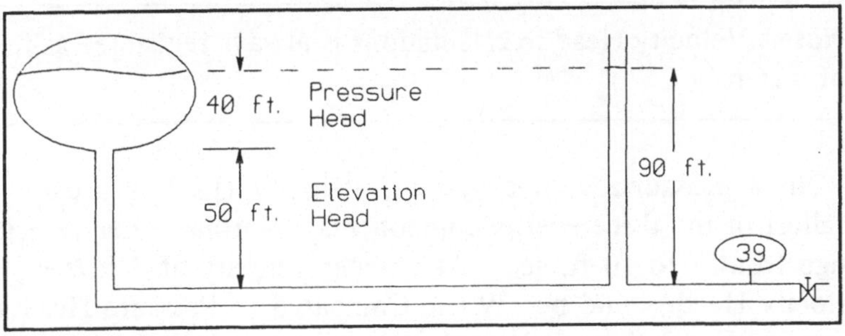

Pressure for community water use can be provided adequately by Pressure Head alone (standpipe), or by a combination of Pressure Head and Elevation Head (elevated tower).

Note: Elevation Head, when used in calculations, is always registered as feet of water. It is often symbolized as Z.

Velocity Head

There is one more energy component in a water system: Velocity Head. Moving water, just as a rolling vehicle, possesses some energy because of its motion. For instance, it takes more energy to push a car from a standstill into motion, than it does to keep it in motion. The car has some energy because it is already moving. The same applies to water which has kinetic energy because of its velocity: Velocity Head. It is the distance which the water can move due to velocity energy. Like Pressure Head and Elevation Head, Velocity Head is also expressed in feet of water.

The formula for Velocity Head calculation is:

$$\text{Velocity Head} = \frac{\text{Velocity}^2}{2 \text{ x gravity}} = \frac{V^2}{2g} = \frac{V^2}{64.4}$$

(The velocity is diminished by the force of gravity, 32ft./sec./sec., as the water travels.)

> **Note:** Velocity Head in calculations is always registered as feet of water.

In a pressurized pipe system, Velocity Head is usually the smallest of the three energy components. Normal water velocities range from 2 to 10 ft./sec. At a water velocity of 5 ft./sec., the Velocity Head would be .39 ft. Compared to Pressure Head and Elevation Head, this can be insignificant, and calculations for it are often omitted. In certain instances, however, it takes on importance. Where water exits freely from the end of a pipe, there is no Pressure Head left, but the water is still moving. The energy keeping it going is its Velocity Head.

Where the flow is in an open channel, an aqueduct, stream, or partially filled sewer pipe, the water is not under pressure. It has only the downward slope of the channel to maintain velocity, and Velocity Head is the energy involved.

The most useful function of Velocity Head calculations in pressurized pipe systems is its potential as a conversion unit for flow calculation. We know that Pressure Head and Elevation head are registered on a pressure gage. Velocity Head is not. Recordable water pressure is transferred directly to the gage, because pressure in the pipe registers upward into the gage, as well as downward and to each side (Pascal's Principle). However,

Velocity Head only registers forward, in the direction that the water is moving, and will not push upward into a pressure gage. A PITOT GAGE, a flow measuring device, is used to capture Velocity Head, and compare it against Pressure Head. The device may have different configurations, but is basically composed of two gages installed as a unit. One of the gages has a bent end which is directed backward into the flow. This one will record the pressure of the water (Pressure Head), plus the Velocity Head. The other gage extends straight into the pipe and registers only water pressure (Pressure Head).

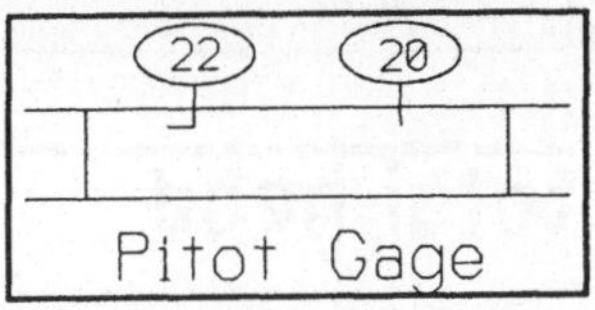

To calculate flow with a pitot gage:

- Convert psi readings from both gages to feet of water.
- Subtract one from the other; this value is the Velocity Head, in feet.
- Convert to water velocity with the Velocity Head formula.
- Use this and the cross sectional area of the pipe to calculate flow (Q = AV).

For practical use in field operations, flow dimensions are set up on charts, based on these differential pressure readings.

Using this method, differential pressure readings enable us to calculate flow. The real importance of Velocity Head in a system under pressure, is not its magnitude as an energy component, but its capacity to become the connecting link to flow calculations.

Note: There is no Velocity Head in a static system. The water is not moving.

If in the system pipe diameter remains the same from point to point, then the velocity is the same at both points, and the Velocity Head is the same at both points.

Note: Velocity Head only changes from one point to another on a pipeline if the diameter of the pipe changes.

total head

The total energy at any one point in a hydraulic system is the sum of the Pressure Head, Elevation Head, and Velocity Head. The summation of these energies is expressed with a formula:

Total head = Pressure Head + Velocity Head + Elevation Head

$$Th = Ph + Vh + Z$$

head loss

When considering water traveling in a dynamic system, some pressure is lost along the length of the pipe because of friction encountered in rubbing against the pipe wall. Turbulence occurs, and energy is lost as head which then dissipates to the surroundings. The amount of loss depends upon system length, diameter, flow, and interior roughness of the pipe. This is Friction Head Loss. It cannot be recovered, and must be calculated in by the utility when considering minimum system pressures to be maintained.

steady state flow

At this point, it should be emphasized that most hydraulic calculations deal with a system in Steady State Flow. This means that the amount of flow crossing a given point does not change with time. We must assume that the flow and pressure source are constant, even though we know that in actual water systems flow varies as water tanks fill and empty.

One important fact becomes apparent in a steady state flow situation. THE FLOW THAT ENTERS THE SYSTEM, EXITS

THE SYSTEM. As long as there are no leaks or lateral exits diverting the water elsewhere, it will not disappear or bunch up inside the pipe (water is incompressible); what goes in must come out. If the pipe diameter changes, the velocity will change, but the flow stays the same ($Q = AV$). In a dynamic system, friction losses will occur, and pressure readings will drop along the length of pipeline, but the flow remains constant. If head losses are suddenly increased, this will cause pressure to drop and then less flow will enter the entire pipe system; but once it has entered, it must exit at the same flow rate.

bernoulli's theorem

Throughout a fluid system, at any one point the total energy is the same as at any other point, as long as friction losses are ignored. That total energy is composed of Pressure Head, Velocity Head, and Elevation Head. If there is a reduction in the energy of one form, there must be an equivalent increase in another. This is Bernoulli's Theorem.

The Total Head formula is applied to a system in which two points are considered:

<u>Point 1</u>		<u>Point 2</u>
$Ph_1 + Vh_1 + Z_1$	$=$	$Ph_2 + Vh_2 + Z_2$

A typical application is that of a pipeline in which the elevations of the two points differ. Draw a Datum through Point 2. A Datum is a reference elevation from which to measure; it is best to draw the datum horizontally through the point of lowest elevation, to avoid the use of negative numbers. The pressure at Point 2 is higher than at Point 1, but the elevation of Point 1 is greater, by exactly the same amount.

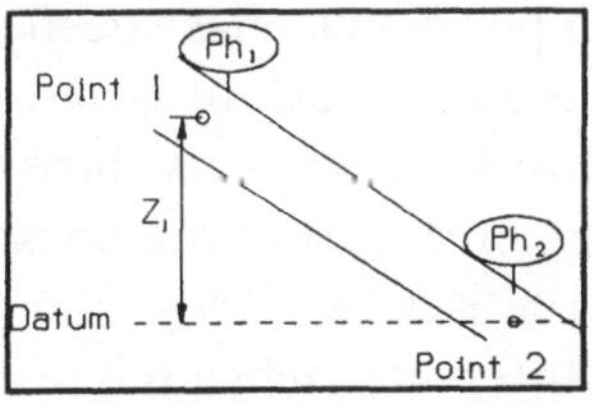

Inspecting the diagram, we can see that the increased Pressure Head that registers on the gage at Point 2 has resulted from the

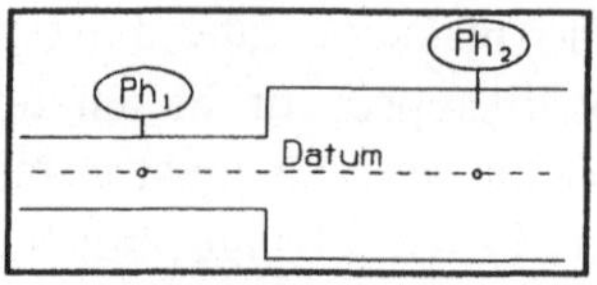

elevation of Point 1. When values are inserted into the formula for each point, the two sides of the equation become equal. Since pipe diameter is the same throughout the system, both Velocity Heads are the same.

Take another example, that of a pipe whose diameter differs from point to point. Point 1 lies in the pipe with the narrower diameter, Point 2 in the wider pipe. Referring back to Q=AV, in a steady state system, if the cross sectional area increases, the velocity of the water will decrease. Therefore, the Velocity Head at Point 2 will also decrease. Bernoulli's formula states that the total energy must remain constant from point to point. That means that one of the other forms of energy must increase at Point 2. Viewing the diagram, we can see that it is not the Elevation Head. This is a horizontal pipe. It is the Pressure Head that increases, and by the amount that the Velocity Head decreased from Point 1.

Note: When velocity decreases, pressure increases.

The opposite would happen if the pipe had decreased in diameter. Velocity increases, and pressure decreases. Utility personnel take advantage of this phenomenon when using the Venturi Meter, another flow measuring device. The structure of a venturi is similar to a pipe which has been squeezed in the middle, creating a narrow diameter at the constricted section. A pressure gage is inserted at this point, and another one just upstream from the constriction. The gage at the constricted throat will read a slightly lower pressure than the one upstream, and the difference, when the readings are converted to feet of pressure, will be the difference in the Velocity Heads between the two sections.

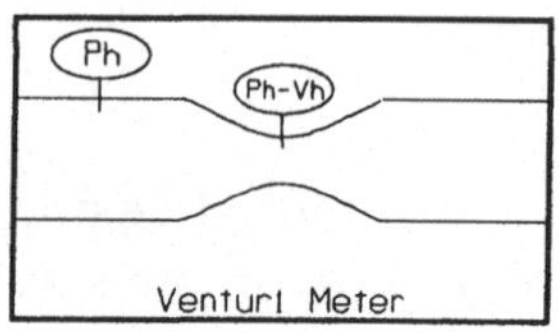

In these situations which produce a pressure differential due to a velocity change, remember that it is the Velocity Head difference that is equivalent to the pressure change (not the velocity).

These internal energy changes within a pipe system are results of structural changes and can be reversed by reverting back to the original structure. Downstream of the constriction, if the pipe enlarges to the original diameter, as it does in the venturi meter, the original pressure and velocity are returned.

Now let us consider real systems, in which there is friction loss. It is a fact that there actually is some energy lost and a drop in pressure from point to point due to pipe friction. We can account for this by adding friction loss into the equation also. If the total energy at Point 1 is not equal to the total energy at Point 2 because there has been some pressure lost due to friction (energy at 2 is less), then we can make it equal by adding the Head Loss to the energy at Point 2.

The formula now becomes:

$$Ph_1 + Vh_1 + Z_1 = Ph_2 + Vh_2 + Z_2 + HL_f$$

As applied to real pipe systems, and as a mathematical formula, the energies at the two points are now equal.

> **Note:** Do not forget - all quantities must be entered as feet of water.

Consider another situation: when the system ends with an open pipe from which the water flows freely. In this case we are viewing the entire pipe system, and all the pressure energy is reduced to zero as the water reaches the end of the pipe. The Velocity

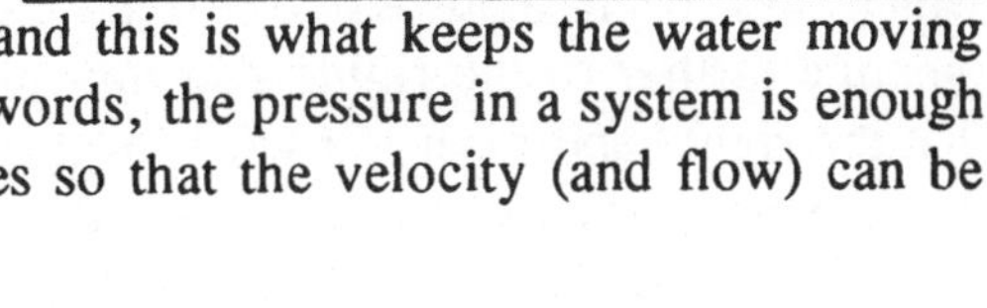

Head remains, however, and this is what keeps the water moving out of the pipe. In other words, the pressure in a system is enough to overcome friction losses so that the velocity (and flow) can be

maintained. The flow admitted into and through the pipe is the amount that will reduce the head loss to zero at the end of the pipe.

If the pressure at the source were greater (deeper), or if the pipe were shorter (less friction loss), then a greater flow would pass through the pipe, and there would be a steeper drop in the Hydraulic Grade Line.

Now picture a valve at the end of this pipe. If it were almost closed, a small flow would pass through the pipe. The Hydraulic Grade Line would drop only slightly along the length of the pipe from pipe friction, until it reached the valve, where it would drop sharply because turbulence at the valve would decrease the pressure almost to zero. In this case flow is sacrificed to maintain pressure in the pipe.

If the entire system were suddenly to become valveless and much shorter, as in the case of a water main break, a great deal more flow would emerge from the end of the pipe. Most of the Head Loss components would have been eliminated, but that system still would have the same amount of source pressure, and all of that pressure must be depleted by the time the water gets to the end of the pipe. Flow would increase enough to increase the head loss in each foot of pipe, so that all the pressure is gone as the water reaches the rupture.

One other situation should be considered at this time. When a hydraulic diagram such as the upper one here depicts one reservoir emptying into another, and points out the differences in water levels between the two, it is meant to bring out the fact that the decreased depth of water in the second reservoir is a direct result of the head loss encountered in the pipe as the water passed from one reservoir to the other. To clarify, observe the lower diagram which has been adjusted to extend the pipeline further to the right; also imagine that the right hand reservoir were much narrower - we can see now that the reservoir (or any water tower) is nothing more than a fat

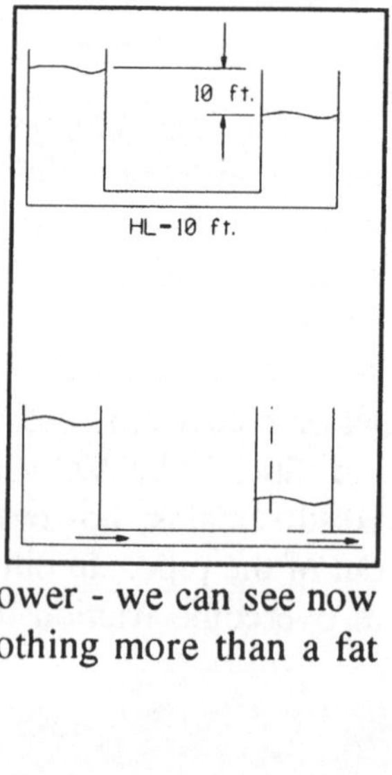

piezometer, and its depth will register the system pressure at that point on the pipeline.

For simple hydraulic situations, such as two points which differ in elevation alone, a quick sketch of the system allows mental calculation for the unknown. The solution would work out just as accurately if all values were inserted into Bernoulli's formula, but this is not always necessary. In more complex instances, however, such as a pipe system which changes in elevation, and pipe diameter, and involves some friction loss, the usefulness of having a well defined formula into which values can be inserted is obvious, in order to avoid mistakes.

Note: The secret to success is: DRAW THE DIAGRAM!

4

pumping - introduction

pumping heads

A pump is installed to move water when existing pressure is inadequate. For instance, water will not move from the treatment plant up into the elevated storage tank on its own; the High Lift pumps provide the needed pressure to lift the water to this height. In determining the amount of pressure a pump must provide to get the job done, the nomenclature of PUMP HEADS was established.

Water enters the pump on the suction side, where the pressure is lower. Since the function of the pump is to add pressure to the system, discharge pressure will always be higher. In pump systems, measurements are taken from the point of reference to the centerline of the pump (horizontal line drawn through the pump). Recall the distinction between:

STATIC WATER SYSTEM - water not moving

DYNAMIC WATER SYSTEM - water moving

Consider a pump which is lifting water from a lower tank to a higher tank:

Static Heads

STATIC SYSTEM - Pump Is OFF!

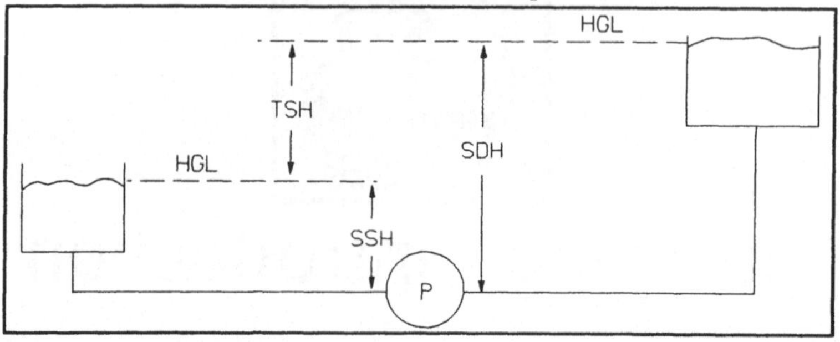

The vertical distance from the water surface on the suction side, to the pump centerline, is the STATIC SUCTION HEAD (SSH). Static Head is always a positive value. It is pressure provided by the system, work the pump does not have to do. Even without the pump, the system would lift water to this height on the right sidc of thc diagram.

Static Suction Head is registered on the gage at the suction side of the pump - when the pump is off.

The vertical distance from the pump centerline to the water level on the discharge side of the pump is the STATIC DISCHARGE HEAD (SDH). This is the total amount of work done by pump and system together. The water was actually lifted this distance. The gage on the discharge side of the pump in a static system will register Static Discharge Head.

The vertical distance between the Static Suction Head and the Static Discharge Head is called TOTAL STATIC HEAD (TSH). This is the theoretical amount of work that the pump must do in lifting water up to the higher tank. To provide this much head is why the pump has been installed.

Pumps are frequently employed to draw water up from below the level of the pump, as in the case of a well. The configuration is such that on the suction side there is a lift, a STATIC SUCTION LIFT (SSL). The pump must have the energy to provide this lift.

STATIC SYSTEM - Pump Is OFF!

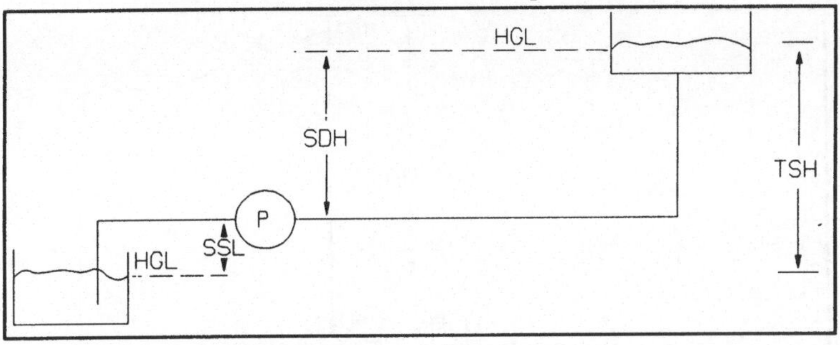

On the discharge side the pump is also providing the head needed to raise the water to the tank above, the STATIC DISCHARGE HEAD. In pump systems which provide a suction lift, the TOTAL STATIC HEAD is the sum of the Static Suction Lift and the Static Discharge Head. The pump is doing all the work.

Dynamic Heads

If a pump system is to provide pressure to move water, obviously the pump must be on. Suction and discharge heads are not static. They are dynamic heads, and they will depend upon the length and condition of the system. Due to friction head loss, suction and discharge heads will decrease with pipe length. The Hydraulic Grade Line will slope downward toward the pump on the suction side, and downward away from the pump on the discharge side, designating that some pressure has been lost because this water is moving; the system is dynamic.

The DYNAMIC SUCTION HEAD (DSH) is the vertical distance between the water level on the suction side and the centerline of the pump, minus the friction headloss.

The DYNAMIC DISCHARGE HEAD (DDH) is the vertical distance between the pump centerline and the water level on the discharge side, plus the friction headloss. The pump must make up this pressure lost in transit in order to get the water up into the tank.

DYNAMIC SYSTEM - Pump Is ON!

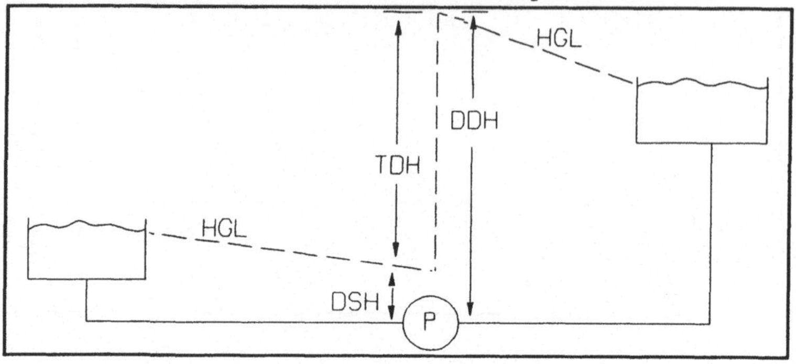

The Dynamic Discharge Head is the pressure that the pump discharge gage will record when the pump is on.

The TOTAL DYNAMIC HEAD (TDH) is the difference between the Dynamic Suction Head and the Dynamic Discharge Head. The actual work this pump must do in lifting water to the higher tank is greater than that designated by the static heads. The pump must make up friction losses on both suction and discharge sides.

Again, look at the suction lift situation. Since pressure is lost in transit to the pump, the DYNAMIC SUCTION LIFT (DSL) is the vertical distance between the water level and the centerline of the pump, plus the friction headloss. It will register as negative pressure (vacuum) on a gage on the suction side of the pump.

The DYNAMIC DISCHARGE HEAD is the vertical distance between the pump centerline and the water level on the discharge side, plus the friction headloss. The TOTAL DYNAMIC HEAD in this system is the sum of the Dynamic Suction Lift and the Dynamic Discharge Head - and it includes friction losses on both sides of the pump. This is the energy needed to move the water; it is this pumping head which determines the size of pump needed for this operation.

DYNAMIC SYSTEM - Pump is ON!

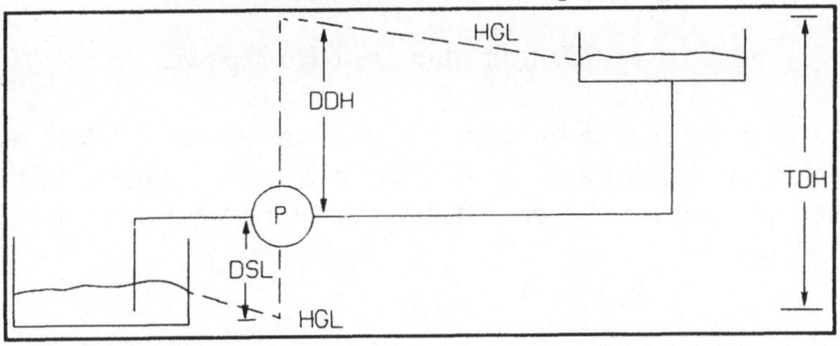

The same pumping heads also can be identified in a booster pump situation. When line pressure available from the source has dropped to too low a level, a booster pump is installed to increase pressure. There is no change in elevation here; the pump is installed solely to restore the pressure depleted by friction losses in the system.

power calculations

In order to calculate the cost of operating a pump, it is necessary to equate the amount of water moved to units of electrical energy consumed in moving that water. In order to purchase a pump, it must first be determined how much work that pump will be required to do.

We must start with the concept of work, the operation of a force over a specific distance. The unit of work is the foot pound: the amount of work required to lift a one pound object one foot off the ground (ft.lb.).

When considering work being done, it is more valuable to be able to do the work faster, and so we consider the rate at which work is being done. This is called Power, and is labelled as foot pound/second. At some point it was determined that the ideal work animal, the horse, could move 550 pounds one foot, in one second. Since large amounts of work are also to be considered, this unit became known as Horsepower.

$$550 \text{ ft.lb./sec.} = 1 \text{ Horsepower}$$

or

$$33{,}000 \text{ ft.lb./min.} = 1 \text{ Horsepower}$$

A pump also does work. It also has power. It lifts water (which has weight) a given distance, in a specific amount of time. The unit label is the same: ft.lb./min. We state it this way:

$$\frac{\text{ft. x gpm}}{3960} = \text{whp}$$

This is the standard formula for horsepower, derived from the amount of lift, or head a pump will be expected to provide, and the flow at which the pump will be operating (pounds is changed to gallons by dividing by 8.34 lb/gal.; 33,000 ft.lb./min/hp is added to yield horsepower units, and in the derivation gets divided by the 8.34, yielding 3960).

This is WATER HORSEPOWER (whp), the power necessary to lift the water to the required height.

The pump does not operate alone. It is driven by the motor, and electrical energy drives the motor. Neither pump nor motor are 100% efficient. There are friction losses within both of these units, and it will take <u>more</u> horsepower applied to the pump to get the required amount of horsepower to move the water, and even <u>more</u> horsepower applied to the motor to get the job done.

BRAKE HORSEPOWER is the horsepower applied to the pump.

MOTOR HORSEPOWER is the horsepower applied to the motor.

EFFICIENCY is the power produced by the unit, divided by the power used in operating the unit.

Water Horsepower must be divided by the pump efficiency to obtain Brake Horsepower.

Brake Horsepower must be divided by the motor efficiency to obtain Motor Horsepower.

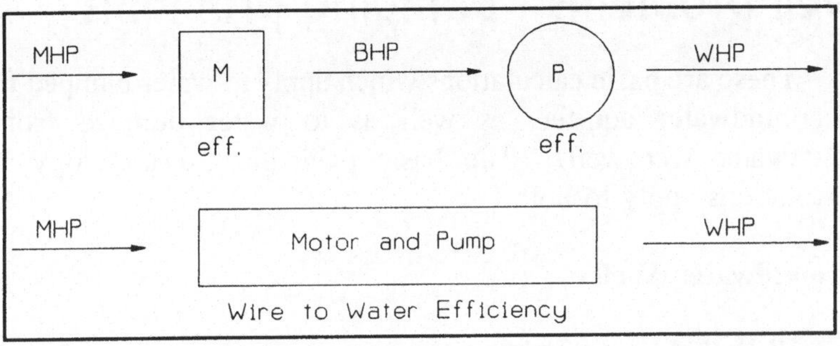

Most pumps operate at 60-85% efficiency. Motors are usually 80-90% efficient.

Multiplying the pump efficiency by the motor efficiency will provide the total system efficiency, the WIRE TO WATER EFFICIENCY.

Next, it is necessary to be able to convert units of horsepower, to units of electrical energy. There is a direct conversion:

1 Horsepower = .746 Kilowatts

This power demand, in Kilowatts, is sold by the hour (Kilowatt Hour). Multiply then, by the number of hours the pump runs, and then by the cost per Kilowatt Hour, and we have the cost of running the pump.

In summary:

$$\frac{\text{gpm x head}}{3960} = \frac{\text{water Hp}}{\text{pump effic.}} = \frac{\text{brake Hp}}{\text{motor effic.}}$$

$$= \text{motor Hp x .746 } \frac{\text{Kw}}{\text{Hp}} \text{ x hrs. x } \frac{\text{cost}}{\text{Kw.hr.}} = \$$$

well problems - pumping problems

These are basic calculations which apply to water pumped from a groundwater aquifer, as well as to water pumped from a wastewater wet well. The basic principles, terminology, and calculations apply to both.

Groundwater Well

Well Yield: the volume of water pumped in a specified time period. It is expressed as gallons per minute (gpm).

Static Water Level: the water level when the pump is off.

Pumping Water Level: the water level when the pump is on.

Drawdown: the difference, measured in feet, between the static water level and the pumping water level. It is the distance the water level drops once pumping begins. The vacuum created at the end of the pump intake line brings water from the immediate area into the well, and a depression is created in the water level in that area.

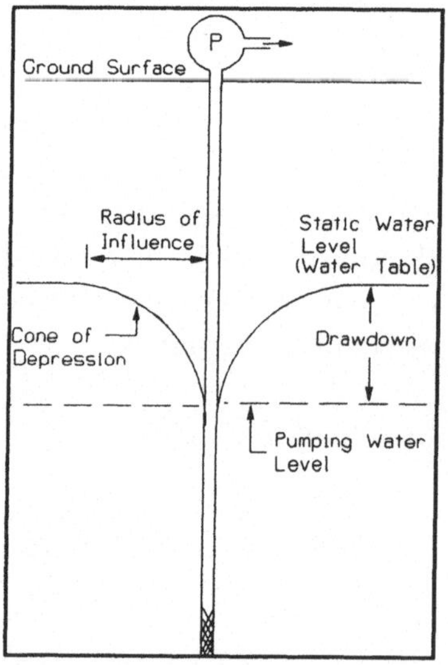

Safe Well Yield: the maximum pumping rate that can be achieved without increasing the drawdown. In a groundwater well, drawdown can easily be fifty feet or more. It may take an hour or so to stabilize and arrive at a permanent pumping water level, but once this is achieved, the <u>drawdown should not be increasing with time</u>. If it is, then the pump is withdrawing more water than the aquifer can return to the well. This is called Water Mining. Eventually the pumping water level will drop below the well screen, and the well will fail. Overpumping an entire aquifer can have permanent effects on the land: sinkhole formation, and in areas near the ocean, saltwater intrusion. Natural phenomena can also influence well supply. Drought lowers the static water level so that even with acceptable drawdown the pumping water level may be too low in dry years for effective pumping. Excessive rainfall raises the static water level, reversing the effect. Soil porosity and geologic structure also influence the aquifer's ability to return groundwater to the well, and limit pump size.

Radius of Influence: the distance between the pump shaft and the outermost area affected by drawdown. This is important in well fields with many pumps. If wells are too close together, and radii of influence overlap, the drawdown will be increased in all the wells.

Cone of Depression: the curve of the line which extends from the pumping water level to the static water level at the outside edge of the radius of influence. Design calculations based on the slope of this curve will determine the radius of influence.

Specific Capacity: pumping rate per foot of drawdown (gpm/ft.). This is a measure of well yield, and is recorded to indicate changes in drawdown and/or pumping rate.

Wet Wells

Water pumped from a tank by a pump set above the water surface exhibits the same phenomena as the groundwater well, except that static and pumping water levels are not separate because there is no soil structure to create drawdown. As the pump draws water up, the water fills back into that space immediately. However, the word "drawdown" is used, and it refers to the distance that the water level is lowered by a pump which withdraws water from a tank. If the pumping rate is exactly the same as the rate at which the tank is filling, the water level in the tank will remain the same.

In water and wastewater applications, this is very rarely the case. Water enters the tank at varying flows depending on the source, and pumping rate must vary to accommodate for large and small flows. Level control devices govern pump operation, and pumping rate can be checked by running a pumping test (pump for a specific time from a well with known volume; record drawdown; calculate gallons of water pumped out; divide by time span of the test). If at the same time water is also flowing into the tank, this must be calculated in too.

water level sensing and pump control

Mechanical Floats: extend from side of tank, riding on water surface. At high and low water levels, a transmitter is mechanically activated and signals the pump.

Diaphragm Element: in a small air chamber, its movement responds to pressure created by water depth above it, and signals the pump.

Bubbler: when high and low level pressure needed to pump air through a narrow tube extending to bottom of tank are reached, pump is activated.

Electronic Probes: installed into tank at known depths, sense water rising over them, and signal pump to action.

Ultrasonic Detectors: transmitter above sends high frequency sound waves down to water surface; the time it takes for the wave to bounce back is proportional to water level.

5

friction loss

This chapter will consider friction loss in closed pipe systems. Recall that when water flows in a pipeline under pressure, the forward movement of the water is a result of the weight of water behind it, pushing it forward. This pressure, registered in feet, is called Head. It does not matter whether the pressure originates from a standing tank of water (Pressure Head), or from the slope of the pipe (Elevation Head) extending down to the point of measurement, or from a combination of the two. In a closed conduit, velocity and flow are dependent upon pressure.

In a dynamic system, pressure decreases along the length of the pipeline; the Hydraulic Grade Line slopes downward, designating pressure loss from friction created as the water encounters the sides of the pipe. This is called Major Head Loss, and usually accounts for most of the pressure drop in dynamic water systems. Minor Head Loss is caused by extra turbulence created at bends, fittings, and diameter changes in the pipeline; it is usually a lesser component of head loss, and will be taken up later in the chapter .

components of head loss

Roughness

The interior surface of a pipe is rough, even when the pipe is new. This roughness is dependent upon pipe material, and will increase with corrosion and age, as the pipe becomes tuberculated and pitted. Normal flow in water pipes is turbulent, and that turbulence increases with pipe roughness; energy is spent, and pressure drops over length.

Length

Friction losses occur with every foot of pipe length, and this must be included in head loss calculations. The longer the pipe, the more head loss.

Diameter

Pipe diameter is also an important factor. In a large diameter pipe, less of the water is actually touching the sides of the pipe and encountering friction, than in a small diameter pipe. Therefore, small diameter pipes have more potential head loss than large diameter pipes.

Velocity

The faster the water flows through a pipe the more turbulence is created, and the more head loss there will be. Therefore, the Velocity Head also contributes to head loss. Recalling $Q=AV$, we can also state: for the same diameter pipe, when flow increases, head loss increases.

In summary, Friction Head Loss (HL_f) is dependent upon:

- material and condition of pipe (roughness)
- length of pipe
- diameter of pipe
- Velocity Head

In the development of formulas to calculate head loss, these very components are used, but with time some specialized terminology has resulted:

C Factor

Also referred to as the Coefficient of Friction, C Factor is a designation of interior pipe roughness, and is set up as a numerical scale which ranges 10-150. Based on experimentation, accepted tables of C factors have been established for pipe, and one is included here. Estimates may be made for pipes not included, and generally C factor decreases by 5-10 with each five years of pipe age.

Note: A high C factor means a smooth pipe; a low C factor means a rough pipe.

Flow for a system is often calculated with a C factor of 100, based on averaging it over the life of the pipe system. A knowledge of the particular system is invaluable here, for pipe roughness is very dependent upon water quality, and historical data on the chemical condition of the particular water and its effect on the interior surfaces of the pipes can determine what value is chosen for C.

Values of C for Hazen-Williams Formula

Type of Pipe	C
Asbestos Cement	140
Brass	140
Brick Sewer	100
Cast Iron	
New, unlined	130
New, lined	140
10 years old	110
20 years old	90
Ductile Iron, (cement lined)	140
Concrete or Concrete Lined	
Smooth, steel forms	140
Wooden forms	120
Rough	110
Copper	140
Fire Hose (rubber lined)	135
Galvanized Iron	120
Glass	140

Values of C for Hazen-Williams Formula (continued)

Type of Pipe	C
Lead	130
Masonry Conduit	130
Plastic	150
Steel	
Coal-tar enamel lined	150
New Unlined	140
Riveted	110
Tin	130
Vitrified	120
Wood Stave	120

Slope

"Slope" as an English word means the amount of slant, or incline. This applies to water systems also. In open channels (aqueducts, sewer pipe) where the water flows by gravity, slope is the amount of incline of the pipe, and is calculated as feet of drop per foot of pipe length (ft./ft.) This slope is designed to be just enough to overcome friction losses, so that the velocity will remain constant, the water will keep flowing, and solids will not settle out along the way. The same principle holds true in a closed pipe system under pressure. We are here considering a horizontal conduit which encounters "feet" of pressure loss for every foot of pipe. The pressure loss would slow down the velocity, restricting flow, and so must be overcome. It is the same as the slope in an open system, but instead of actually slanting the pipe to overcome the losses, a source of pressure is added to overcome friction. It is still called "slope": feet of head loss per foot of pipe (ft./ft.) which must be overcome. To calculate, divide pressure loss by length of pipe:

$$s = HL_f/L$$

A widely used formula for head loss calculation was developed by Hazen and Williams:

$$Q = .435\ C\ d^{2.63}\ s^{.54}$$

Its components originated from the simple concepts of roughness (C factor), length, diameter, and Velocity Head (mathematically converted to flow).

With a calculator it is easy to compute, and all components can be measured or estimated directly.

There are many variations of the Hazen-Williams formula for head loss calculation. Some have different exponents; some equate to velocity instead of to flow; some are specialized for open channel flow. Other hydraulics experts have developed their own variations, and use them under their own names, but it is easy to recognize the general format of the equation, and they all calculate out to yield the same approximate value of any one dimension. It is important to stress "approximate", for a few inherent weaknesses exist in any of these formulas.

There is no problem with flow, velocity, slope or length. If these are to be the known factors, they can be measured, metered, read from gages. The weakest point in the formula is the Roughness Coefficient, C. If this is to be a known factor for calculation, at best it can only be estimated, based on pipe type, age, and knowledge of water quality. The only way to accurately determine the amount of roughness in a pipe is to measure all the other components of the formula, and then calculate for C. This is done by utilities to determine when pipe needs to be cleaned.

Pipe diameter can also lead to inaccurate calculations. Most important is knowledge of the particular system's water quality. How much rust buildup, chemical post-precipitation, or scale formation have occurred since that pipe was laid? A six inch diameter pipe can quickly become a two inch diameter pipe if the water quality is poor. In how many systems can we assume that the diameter of the pipe is the same as when it was new? This is where the operator's knowledge of his treatment system counts.

Hazen-Williams Alignment Chart

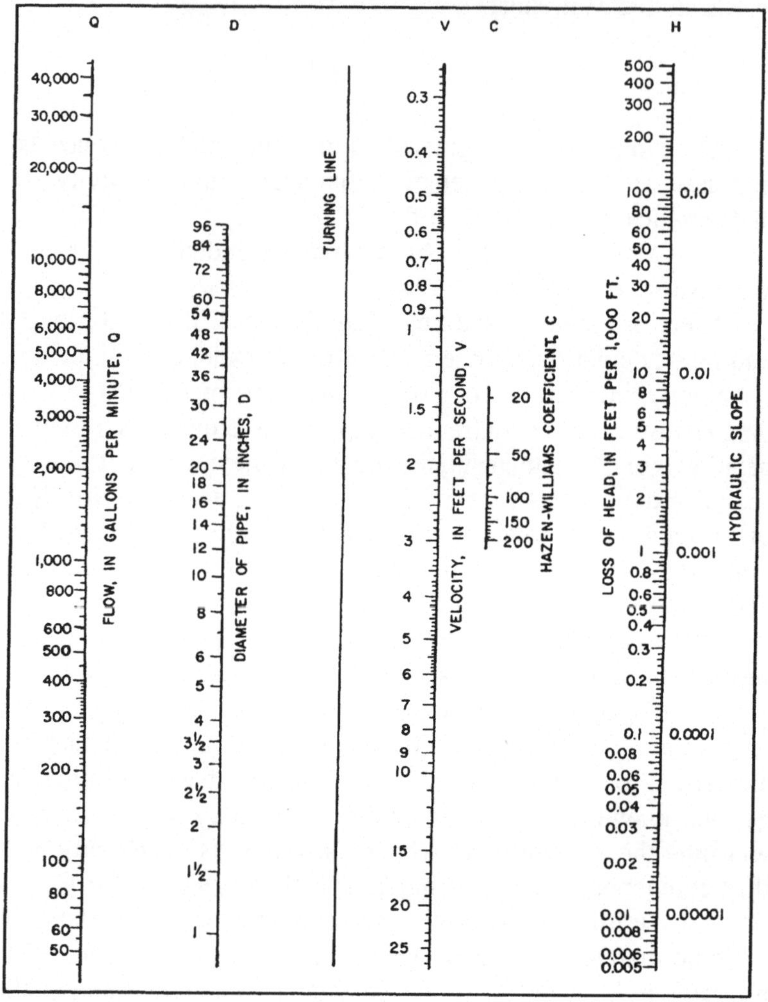

Directions For Using Hazen-Williams Alignment Chart

For determining head loss, find the proper discharge and line it up with the pipe diameter. Extend the line across to the Pivot Line, and mark the point where the Pivot Line was crossed. Line up that point with the proper C factor on the other side, and extend it to the head loss.

Velocity is determined from a line which passes through flow and diameter. Pivot Line is not needed for this. Pivot line is needed only for finding C factor or head loss.

For field use, as an alternate to calculating the Hazen-Williams formula, an alignment chart has become popular, and can be used with reasonable accuracy. This chart is of a special type called a nomograph. It may yield an answer with one or more of its dimensions unknown. Inspecting it, we note dimensions of discharge, pipe diameter, water velocity, C factor, and head loss in ft./1000 ft. There is also a Pivot Line, for getting from one side to the other.

compound pipes

Municipal water distribution systems are not built with separate single pipes extending from treatment plant to customer's residence. Rather, a network of pipes is laid under the streets - different diameters, lengths, ages and materials. Some branch and loop; some are connected end to end. These more complex arrangements are referred to as Compound Pipe Systems. There are two basic types:

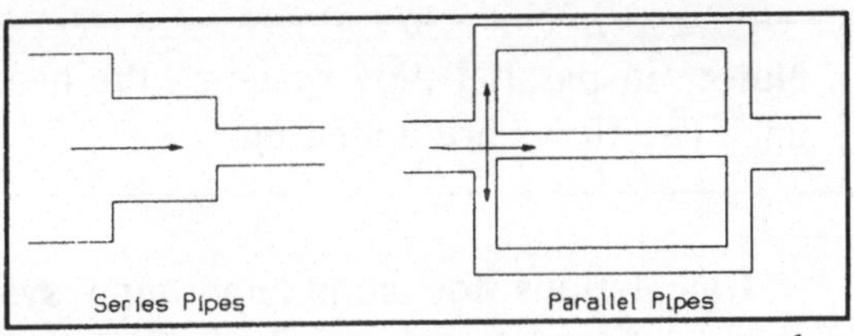

PIPES IN SERIES - two or more different pipes, laid end to end.

PIPES IN PARALLEL - two or more pipes laid side by side, with the flow splitting among them.

Calculating With Pipes in Series

When pipes are laid in series:

A) The same flow passes through all of them
B) The total head loss is the sum of the head losses of the component pipes.

Calculating With Pipes in Parallel

When pipes are laid in parallel:

A) The flow splits among them. Least flow is admitted to the pipe with the most resistance (smaller diameter, longer, rougher). The greatest flow will pass through the pipe with the least resistance (wider, shorter, smoother).
B) The head loss in each pipe component of a parallel system - is the same. The flow split makes it so. Taking the path of least resistance, just enough flow is admitted to each pipe, so that the head losses are balanced.
C) The head loss through the entire parallel system is the same as the head loss through any one of its pipe components.

Note: In parallel pipe systems, the head losses are not added up. The flows are added up.

Calculations for compound pipe systems are the basis for solving piping network problems, and are complex. For example, an analysis of flows and pipe conditions in an entire section of a city water distribution system may be required for planning system expansions, maintenance, major replacements, and for troubleshooting low pressure problems. It is not within the realm of this text to attempt such calculations, and in practice computer systems which automate the process are in widespread use.

minor head loss

To this point we have considered Major Head Loss, the pressure drop resulting from friction along a length of pipeline. Minor Head Loss is caused by turbulence created at bends, valves, fittings, and changes in pipe diameter. It is almost always less than Major Head Loss, and its inclusion in calculations is optional. A city water distribution system has valves and laterals at frequent intervals, and an approximation of minor loss would have to be made to determine its significance. A general guideline can be used: if the minor head loss is less than five percent of the total head loss, ignore it.

equivalent pipe technique

Determination of minor head loss is simple and straightforward, and a standard nomograph in common use is included here. For calculations, each fitting is converted to a length of straight pipe of the same diameter which would cause the same head loss. This makes further calculation easy. Once the fitting has merely been converted to an extra length of pipe, calculations for the total length will include both major and minor losses.

Inspecting the nomograph, note the difference in equivalent pipe length between a fully open gate valve, and a 3/4 closed gate valve. In pumping operations, it is common practice to control flow rate by partially closing a valve on the pump discharge, thereby increasing the minor head loss, and restricting the flow. Using another example, a pressure regulating valve senses a pressure higher than its setpoint on the downstream side of the valve, and closes slightly to increase the minor head loss, thus dropping the pressure. In water distribution systems, we can readily understand the importance of proper valve maintenance and record keeping. A shutoff valve in a large main left half closed by accident would significantly affect the line pressure.

Note the diagram shown for "Ordinary Entrance". This refers to an entrance to the pipe. Similarly "Ordinary Exit" would be an exit from the pipe.

The odd configuration called "Borda Entrance" is most frequently seen as the intake end of a pump drawing a suction lift.

For enlargements and contractions, measure with the smaller pipe size on the nomograph.

Resistance to Flow from Valves and Fittings

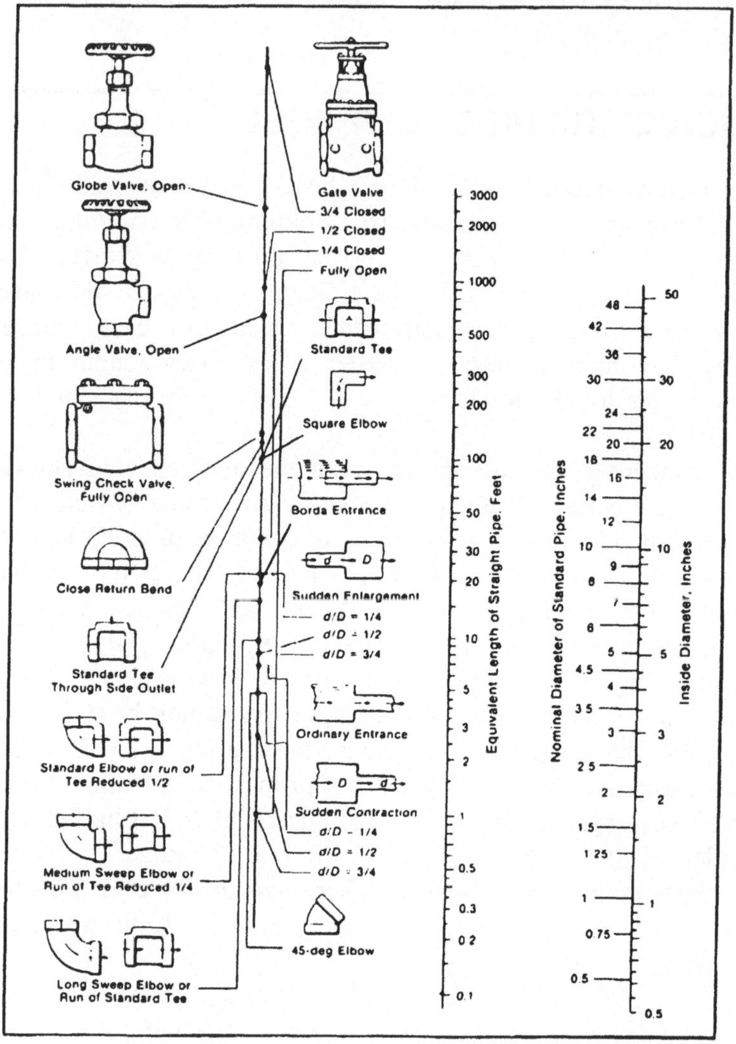

Reprinted with permission of Crane Co.

Two separate hydraulic phenomena are occurring in these areas of pipe enlargements and contractions:

1. Pressure drops because of extra turbulence encountered in passage to a pipe of different diameter (minor head loss). This is a permanent loss of pressure, which must be made up by the source.

2. Pressure and velocity changes are encountered because the pipe has changed diameter. These changes would be only temporary if downstream the pipe returned to the original diameter, and the only pressure loss would then be from major and minor head loss through this section.

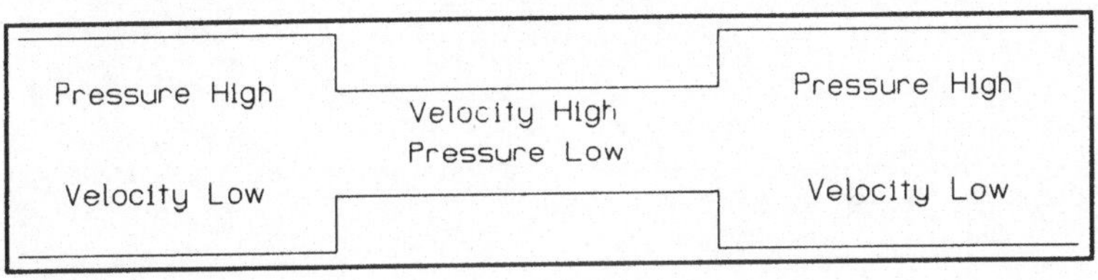

When faced with a fitting not listed on the nomograph estimate. Use the closest one.

Note: Use care with reading nomographs; misplacing the ruler by a fraction of an inch may change the reading by a great deal.

velocity head technique

There are a few minor losses which may be expressed in terms of Velocity Head ($V^2/2g$). These have been worked out by experimentation and measurement, and are useful in calculations where Bernoulli's Theorem is in use and a Velocity Head value is readily at hand. This method directly converts to a head loss value for the fixture.

Ordinary Exit: Head Loss = 1 Velocity Head

Ordinary Entrance: Head Loss = 1/2 Velocity Head

Borda Entrance: Head Loss = 1 Velocity Head

Most often, conversion to equivalent pipe length is the method of choice. Calculations from previous chapters remain the same when minor head loss is included. The only difference is that with this inclusion, the pipe length has been extended.

6

open channel flow

Rivers, streams, aqueducts, drainage and irrigation ditches, and sewer pipes most often flow only partially full. Open Channel Flow exists in these conduits where the water has a free surface exposed to atmospheric pressure. Water depth being constant from one point to another, there is no pressure to act as the driving force. The only natural energy this water possesses is its Velocity Head, and at normal water velocities this is a small value. The basic hydraulic principles apply, however. Open channels have a degree of roughness, just as water pipes under pressure do, and friction occurs as the moving water rubs against the sides of the channel. This quickly depletes the Velocity Head, and decreases the velocity. To cope with this, open channel systems (aqueducts, sewers, drains) are constructed with a slope; the pipe is physically slanted downward just enough to overcome friction losses, and keep the velocity (and flow) constant. As long as the slope of the channel (ft./ft.) is exactly the same as the friction loss (ft./ft.), velocity will be maintained. Man here is making use of gravity to overcome friction, so that the Velocity Head can keep the water moving.

In a properly sloped channel, at steady state flow, the water surface is parallel to the channel bottom. Energy loss has been negated by the slope of the channel, and the depth of the water remains the same throughout. The slope of the channel is overcoming the friction.

If the channel were horizontal, however, the water would decrease in velocity as it encounters friction, and would "pile up" behind, making its own slope. The water produces for itself the Pressure Head needed to keep moving, as a greater depth at the upstream end, and the slope of the water surface will register its progressive loss downstream. The Hydraulic Grade Line is drawn parallel to the water surface, and it registers the loss of energy.

If this same channel were taken and slanted downward at exactly the same slope as the water surface registers, the surface would flatten out and become parallel with the conduit bottom again.

If the channel is sloped more than is necessary to overcome friction, velocity will increase, and water depth will decrease. The extra energy created by the steep slope becomes velocity energy.

If, in an existing channel, the flow increases, the area of flow will increase (water will get deeper), thus holding the velocity constant. Note that in each of these situations, there is a natural adjustment of the water depth, resulting in a change in the cross sectional area of flow.

In wastewater collection systems, if the slope between manholes is not sufficient to maintain scouring velocity, solids will settle along the length of the pipe.

In natural open channels, the same phenomena are evidenced. A stream bed which slopes sharply, then levels off somewhat, will carry water at a shallow depth where the slope is steep, and velocity is high. Downstream, where the slope is gradual, the water will be deeper, and the velocity slower.

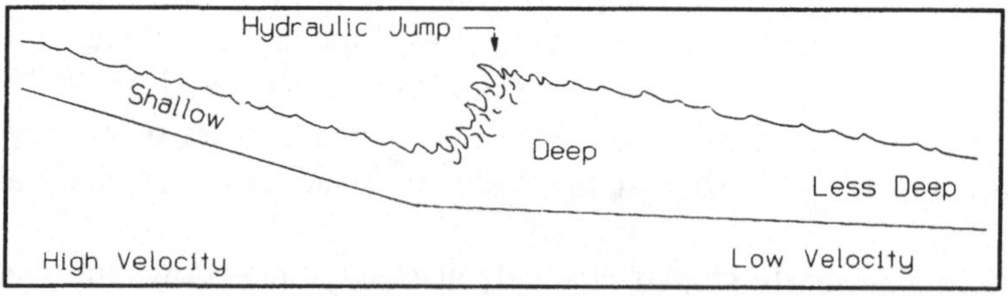

It is interesting to note that in places where waters of two different velocities meet, a short section of deeper water occurs; the water level rises at the point where the velocity changes, and a "hump" forms here, before the surface smoothes out again. There is extra turbulence at this point; the water traps air and expands somewhat. Called Hydraulic Jump, this phenomenon occurs dramatically at dams and flumes, and much less dramatically as shoreline waves or those created from the wake of a boat. On a small scale it is visible as ripples that occur when a stone is thrown into water, or when water pours from a faucet into a full sink or bucket.

A distinction in flow characteristics occurs between open channel flow and closed pipe flow. In a pressurized pipe, the upstream pressure determines the velocity with which the water will travel, and that velocity determines the head loss. The amount of water admitted to the system (the flow) is dependent upon this velocity and pipe diameter. In open channels, however, the head loss is equal to the elevation change produced by the slope of the pipe; this will determine the velocity and the depth of water in the pipe, and this velocity and depth will vary depending upon the adequacy of the pipe slope to overcome friction losses. The total amount of flow approaching the system is admitted however, and the cross sectional area of flow along with the water velocity will adjust to accommodate this flow (unless the channel isn't deep enough, and then the water will flood out over both sides).

calculations for open channel flow

In open channels, maintenance of proper velocity is of prime importance. In order to keep the velocity constant, the channel slope must be adequate to overcome friction losses. Calculation for head loss at a given flow is necessary, and the Hazen-Williams formula is useful.

$$Q = .435 \ C \ d^{2.63} \ s^{.54}$$

The concept of slope has not changed. The only difference is that now we are measuring the physical slope of a channel (ft./ft.), equivalent to head loss.

The problem that arises is with the diameter. In a conduit which is not circular (rivers, grit chambers), or in pipes only partially full (sewers, drains), where the cross sectional area of the water is not circular, there is no diameter. Another parameter must be used to designate the size of the cross section, and the amount of it that contacts the sides of the conduit. For these non-circular areas, we use Hydraulic Radius (R). It is a measure of the efficiency with which the conduit can transmit water, and is determined by dividing the cross sectional area of the water, by the perimeter of conduit in contact with that water.

$$R = \frac{\text{wetted area}}{\text{wetted perimeter}}$$

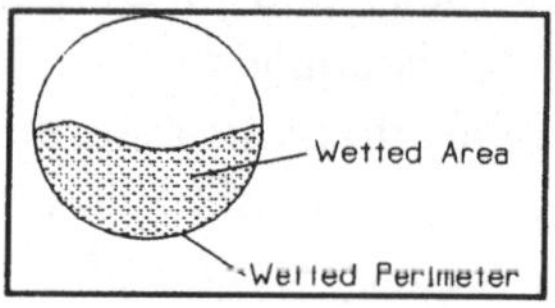

This is an attempt to measure how much of the water is in contact with the sides of the channel - or - how much of the water is not in contact with the sides, the "hydraulic" radius.

For use in open channels, Manning's Formula has become most popular:

$$V = \frac{1.486}{n} R^{.66} s^{.5}$$

This is a variation of the Hazen-Williams formula, where diameter has been replaced by R, thus making the formula flexible enough to adapt to all cross sectional areas.

There is a new component - n. This is a roughness coefficient (similar to C). Materials of which open channels are constructed can be very different from those used in closed pipe flow, especially when we consider that nature has made many of them. Man-made drainage channels are often lined with concrete, asphalt, rip-rap, or grass to prevent erosion from the flowing water. Sewer pipes can be lined or unlined, and are constructed of iron, steel,

Manning Roughness Coefficient, n

Type of Conduit	n
Pipe:	
Cast Iron, coated	0.012 - 0.014
Cast Iron, uncoated	0.013 - 0.015
Wrought Iron, galvanized	0.015 - 0.017
Wrought Iron, black	0.012 - 0.015
Steel, riveted and spiral	0.015 - 0.017
Corrugated	0.021 - 0.026
Wood Stave	0.012 - 0.013
Cement Surface	0.010 - 0.013
Concrete	0.012 - 0.017
Vitrified	0.013 - 0.015
Clay, drainage tile	0.012 - 0.014
Lined Channels:	
Metal, smooth semicircular	0.011 - 0.015
Metal, corrugated	0.023 - 0.025
Wood, planed	0.010 - 0.015
Wood, unplaned	0.011 - 0.015
Cement Lined	0.010 - 0.013
Concrete	0.014 - 0.016
Cement Rubble	0.017 - 0.030
Grass	----- - 0.200
Unlined Channels:	
Earth: straight and uniform	0.017 - 0.025
Earth: dredged	0.025 - 0.033
Earth: winding	0.023 - 0.030
Earth: stony	0.025 - 0.040
Rock: smooth and uniform	0.025 - 0.035
Rock: jagged and irregular	0.035 - 0.045

concrete, brick, stones, clay and many other materials. Manning's n factors developed as a method to approximate roughness in open channels.

Note: n factors decrease as the channels get smoother.

Manning's formula equates to velocity, instead of to flow. Perhaps this originated from the importance of providing scouring velocity in conduits designed to carry solids-laden water, and from the need to determine a slope that would provide that velocity. Solids allowed to deposit along the length of the channel will change the velocity, and eventually block the channel.

It must be kept in mind that although our calculations are based on steady state flow, the actual discharge into these channels is almost never constant, and varies over the course of a day, week, and with the seasons. Most sanitary sewers are designed so that they are half full at average daily flow. Storm drainage channels are not in use at all most of the time, yet must have extremely high capacity during wet weather. This makes design quite a challenge, and operation is not often at design efficiency.

If desired, to save time and eliminate some calculations in dealing with Manning's formula, a nomograph is provided.

Manning's Nomograph

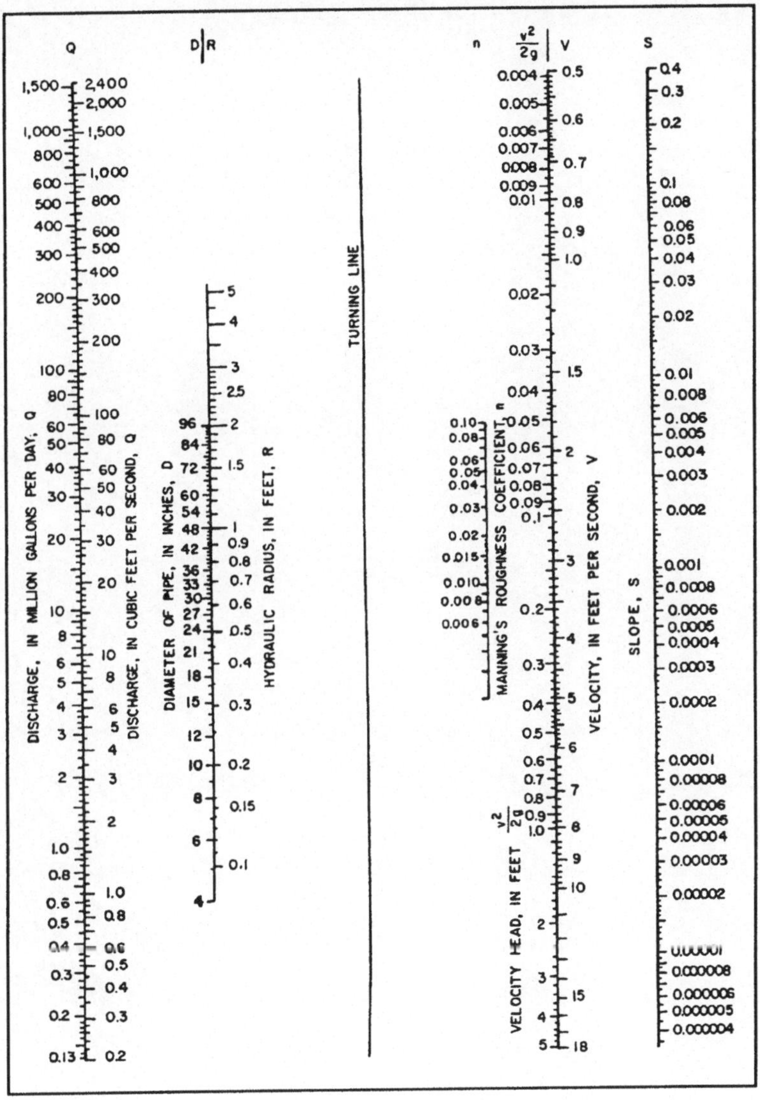

7

flow measurement

Why measure flow? The most vital activities in the operation of water and wastewater treatment plants are dependent on a knowledge of how much water is being processed.

The municipal water works must be able to measure water made and water used in order to calculate percent lost water. Knowledge of flow is needed to size storage tanks and provide firefighting capacity, to guard against backflow, to predict future usage, and to manage water conservation programs.

Wastewater treatment plants need a flow measurement to determine solids and BOD loading to the plant, and to the receiving stream for NPDES compliance. Control of recycle rates, calculation of F/M ratio, and prediction of physical and biological unit performance is based on flow. Knowledge of incoming water quantity is vital to anticipation of plant reaction to industrial input, shock loads, and stormwater. The sizing of stormwater retention basins to prevent flooding and backup is based on a flow reading. The smooth operation of an Industrial Pretreatment Program depends on knowing the quantity of industrial input and how it compares to the quantity of the municipal stream.

Both water and wastewater utilities need flow measurements to calculate chemical feed rates, hydraulic loading to treatment units, tank detention times. It is necessary for scheduling of system maintenance, for calculation of process expenses (power, chemicals, labor), for sizing of pipes, pumps and meters. Flow records are needed to plan treatment plant expansions, to be prepared for legal proceedings, and of course, for customer billing.

Flow measurement can be based on flow rate, or flow amount:

Flow Rate (gpm, MGD, cfs): Utilities need this type of meter inside the treatment plant, in wastewater collection, and in potable water distribution, to determine process variables. In common use are pressure differential meters, magnetic meters, and ultrasonic meters. Some are designed for metering flow in closed pipe systems; some are meant for open channel flow.

Flow Amount (gal., cu.ft.): The totalizer sums up the gallons or cubic feet which pass through the meter. Private, commercial and industrial customer service meters operate this way, and total amount of flow is the basis for billing. In wastewater treatment, automatic composite sampling units are usually flow proportioned to grab a sample every so many gallons. These totalizer meters may be turbine, propeller, compound or positive displacement.

flow rate meters

pressure differential meters: closed pipe flow

Orifice Meters

Orifice meters have been in use since ancient Roman times in water pipes. The principle of operation is that a pressure differential is created across the unit as a steady flow of water passes through a constricted section. We recall that as water flows from an area of large cross section to an area of small cross section, the velocity increases (Q=AV). Since this results in an increase in Velocity

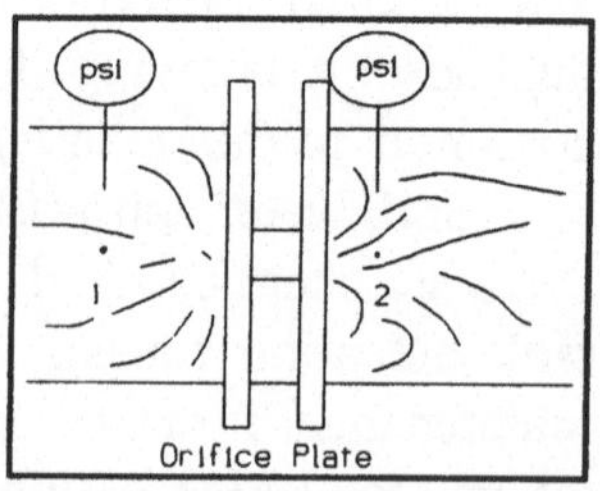

Head, and total energy must be maintained, some form of energy must decrease by the same amount. It is the Pressure Head that drops abruptly at this point. We apply this in practice to measure flow. The sensing unit of the orifice meter is a flat metal plate, usually stainless steel, installed between flanges in a pipeline, with a precisely sized small diameter hole at the center. Pressure gages are installed one pipe diameter upstream, (ahead of turbulence at the plate), and one half pipe diameter downstream (where the jet is narrowest and pressure is minimal) to measure the pressure differential. The formula which calculates flow for an orifice meter is merely a rearrangement of Bernoulli's Theory, and this version has been in common use:

$$Q = C_d \, A \sqrt{(Ph_1 - Ph_2) \times 2g}$$

- The variable A (area) stands for the area of the orifice.
- Ph1 is the Pressure Head upstream of the orifice.
- Ph2 is the Pressure Head downstream of the orifice.

Note: Ph_1 and Ph_2 must be entered as feet of water.

Friction loss is incorporated as a Coefficient of Discharge (C_d), which takes in the loss of head encountered from passing through the meter. It is a fractional value which represents a percent reduction in flow caused by turbulence at the constriction. The value for a sharp (new) orifice is .6; for a round (worn) orifice it is .9.

Note that the pressure drop resulting from the velocity change at the constriction is recoverable as the water continues on the discharge side of the orifice, but the head lost from turbulence created in getting through the constricted section is not.

Other useful applications of this flow formula occur wherever there is an opening or hole in a tank or pipeline. These are orifices, and flow can be calculated. The formula applies just as well in cases where water flows freely from the structure to the atmosphere. In this situation, $Ph_2 = 0$; Ph_1 in the tank or pipe, is the pressure at that point.

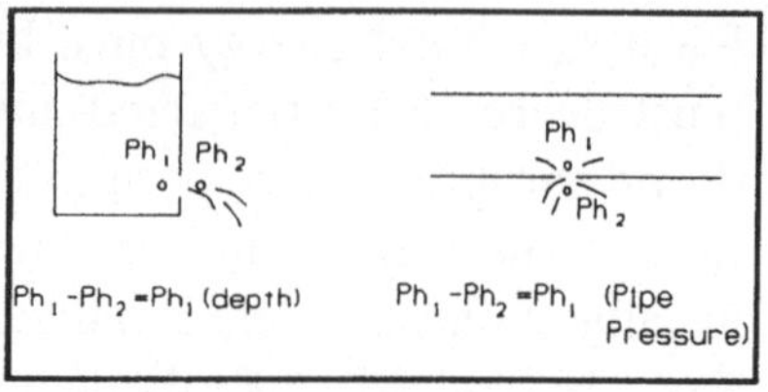

Orifice meters are inexpensive and easy to install. They require little space, and are widely used in industry for water, steam and gas flow measurements. However there are some important disadvantages. Orifice meters are not extremely accurate. The device causes a great amount of turbulence on discharging from the orifice which results in large head loss to the whole system. Note that the presence of the Coefficient of Discharge in the formula could decrease the flow by 40%. Small initial cost must be weighed against cost of pumping, and if high pressure and specific flow quantity are important to the system, this may not be the metering device of choice.

Orifice meters are not recommended for permanent installation to measure wastewater flows; solids in the water easily catch on the orifice, throwing off accuracy.

For installation, it is necessary to have ten diameters of straight pipe ahead of the meter to create a smooth flow pattern, and five diameters of straight pipe on the discharge side. In some installations where a meter is desired, this is not possible.

Venturi Meters

Venturi meters, like orifice meters, measure flow in closed conduits by recording a pressure differential across a constriction in the pipe. The meter consists of a constricted tube, complete with pressure gages or manometer, and is purchased as a unit. It has a converging section, a throat (where water velocity increases), and a longer diverging section. The gages are placed just upstream from the convergence, and at the throat. The pattern of flow is smooth through the unit, and there is very little pressure loss due

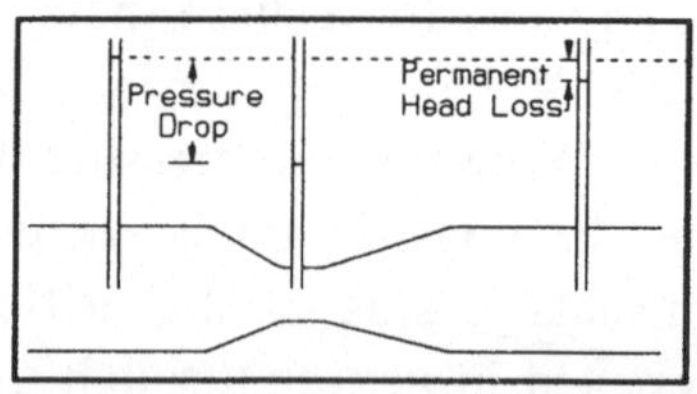

to turbulence. The angle of divergence is about 7%, which does require the tube to have substantial length. Like the orifice, the flow formula for venturi meters is derived from Bernoulli, and it was he, in the 1700's, who first described its operation. Calculation is basically the same as for the orifice, and the orifice formula will yield a good approximation of flow rate. The area used is the area of the throat.

The venturi meter is the most accurate and the most popular closed conduit differential pressure meter. It can be used over a wide range of flows, and has a low permanent head loss because of the smooth configuration. It is applicable to the measurement of wastewaters and sludges, for there are no protruding parts to catch solids or wear rapidly. Maintenance is a matter of flushing and cleaning solids out of the pressure gage tubes. Calibration is done by testing at a known flow rate.

Disadvantages of the venturi are its long length and high capital and installation costs.

Flow Rate Controller

In water filtration facilities which operate constant rate filters, the rate of flow through the filters is regulated by a valve/venturi mechanism, the Flow Rate Controller. It is installed in the filter effluent pipe, and responds to flow changes through the filter sand. As the pores in the filter media fill up with suspended matter, head loss in the bed increases, and less water flows through. At the beginning of the filter run, the controller valve is almost shut. Water passing through the filter has little resistance, and flow would be high, except that passing through this almost-closed valve it encounters enough head loss to drop the rate to a predetermined set point. As the filter bed clogs, passing less and less flow through, the controller valve responds to the dropping pressure from the venturi, and gradually opens, thus keeping the flow constant.

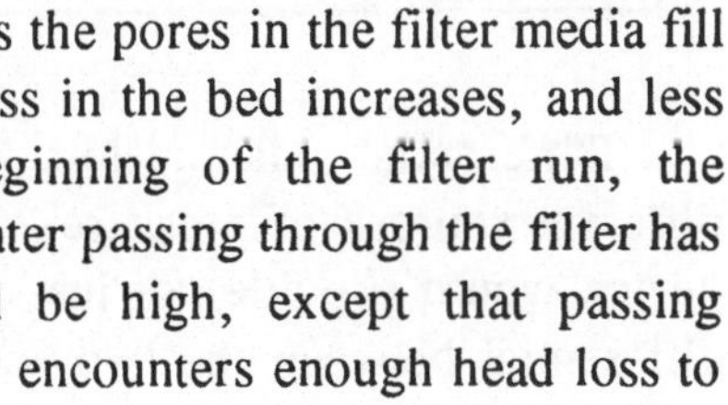

Differential Manometers

Manometers can be installed in venturi meters or orifices instead of the pair of pressure gages. A manometer is a piezometer tube curved into a U shape, and filled with a liquid of known Specific Gravity. Pressure on one end of the tube causes the liquid to move, by an amount equivalent to the pressure.

Simple Manometers are connected to the pressure source at one end, and the internal liquid is free to ride up the other end, as the pressure moves it. These simply read pressure. Manometers installed on anaerobic digester gas lines are of this type. They are also often seen installed on liquid chemical storage tanks, and are used as a means to determine liquid level. The internal fluid may be brightly colored for visibility. The barometer, a manometer which responds to air pressure, is of this type also. It is in contact with the atmosphere at one end (the pressure source), and the free end is enclosed in vacuum. The unit is placed up against a calibration scale so that the liquid movement can be measured.

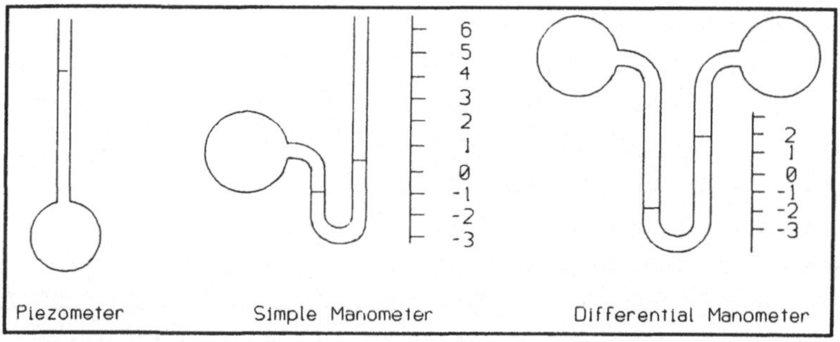

Differential Manometers are connected at both ends, to different sources of pressure. Installed into a venturi meter, the device would provide not just a pressure reading, but a pressure differential between the two sources. The center liquid can ride back and forth depending on which side has the greater pressure. By attaching the differential manometer to both high and low pressure sides of the venturi constriction, the difference in pressure can be measured, and then converted to flow rate. Installed this way onto water pipes, the pressure source (water) must be kept separate from the center fluid by an impermeable membrane.

The two most commonly used center fluids in differential manometers are water and mercury. Water is used as the fluid (colored for visibility) when the pressure differential is expected to be less than 1.25 psi. If the pressure difference were 1 psi, for instance, it would move the internal fluid in a water manometer 2.31 ft. Anything much larger than this would be unwieldy.

Mercury is most often used in manometers as the center fluid. It is 13.6 times heavier than water (S.G.=13.6) and it takes that much more force to produce the same amount of movement as in a water manometer. Thus, with a significant pressure differential, the fluid in mercury manometers only moves a matter of inches. This is much more readable.

To calibrate a manometer, start with both ends connected to exactly the same pressure (or disconnect it completely) and let the center fluid ride to the middle of the tube, so that it rises to the same level on both sides. Mark that as the zero point on the calibrated scale. Now connect it up to the two pressure sources, and read the rise or drop in fluid level. If the center liquid is mercury, inches of movement on the scale will be read. Convert it to feet. This is now feet of mercury. Convert it to feet of water (multiply by 13.6, the Specific Gravity of mercury). To solve for flow through the unit, insert it into the orifice meter formula. This differential pressure reading, in feet of water, is (Ph_1 - Ph_2). If the center fluid is water, read the differential, in inches. Convert it to feet of water. The Specific Gravity of water is 1, so there is no need to adjust for this. If the center fluid is another substance, multiply by the Specific Gravity of that substance.

pressure differential meters: open channel flow

Weirs

To measure flow in sewers, streams, drains - open channels - weirs have become most popular. They are merely flat or notched metal obstructions with sharp top edges, inserted across the line of flow. The water builds up behind the height of the weir, flows over, and the depth of the overflow water is measured. The depth of water going over the weir is referred to as the "head" on the weir. Since it will be deeper at greater flows, calculations can equate it to flow, and shortened formulas for flow over weirs of specific shapes have been developed.

The two most commonly used weir types are:

Rectangular Weir: for large flows.

$$Q = 3.33 \; L \; h^{1.5}$$

L = width of weir
h = head on weir (measured from edge of weir in contact with the water, up to the water surface).

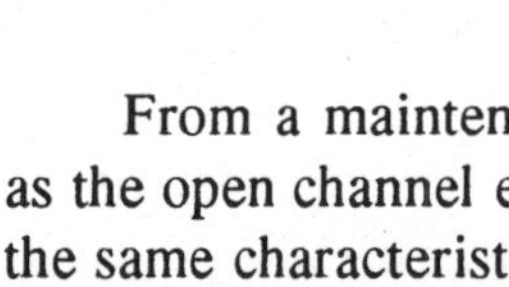

90 Degree V Notch Weir: for small flows.

$$Q = 2.5 \; h^{2.5}$$

h = head on weir (measured from bottom of notch to water surface).

From a maintenance point of view, weirs may be thought of as the open channel equivalent of orifice meters; they have some of the same characteristics. Inexpensive to purchase or make of sheet metal with a wooden backup board for strength, they are easy to install and remove, are adaptable to a wide range of flows, and are very commonly used for temporary measurements of stormflows. They can be permanently installed, but solids carried by the water will catch on the lip or V notch, and decrease measurement accuracy. Frequent maintenance must be planned where weirs are in use.

A depth gage (calibrated stick) is usually used to take the head measurement. In permanent installations, electronic level sensors can be used. Water flowing over a weir contracts at the point of overflow, and the head measurement must be taken just upstream of that contraction. Water flowing over a V notch weir must stay within the notch for accurate measurement. If it doesn't, a larger V notch, or a rectangular weir should be used. Water flowing over any weir must fall free of the weir plate (not dribble down the side) for good accuracy. It doesn't fall free, a weir with a smaller notch should be used.

Weirs are often seen used in clarifiers, grit chambers, and other treatment plant units to reduce velocity or control depth in the unit. Saw toothed weirs, consisting of many small 90 degree V notches, are used as effluent structures in clarifiers. The flow leaving the clarifier is divided by the total number of notches on the weir to determine the discharge per notch. Head on the weir can be calculated from this.

Installing a weir in a channel will effect deeper water (the height of the weir plus the head on the weir) immediately upstream of the weir.

Weirs installed in sewers and manholes may cause backup problems, especially at times when stormwater enters the system. In addition, the extra depth created behind the weir produces a greater cross sectional area of flow, decreasing velocity, in effect to zero, and solids will settle and collect there, throwing off measurement accuracy.

Parshall Flumes

The flume is similar to the venturi meter, except that it is meant to measure flow in open channels. The Parshall Flume is the most popular type, and is much used for the measurement of wastewater and irrigation water, due to its low permanent head loss and self cleansing capacity. Its geometry is smooth flowing; solids will not get caught on it, and it operates well at both high and low flows. The Parshall Flume has three sections:

Inlet: with rapidly converging walls, and a level bottom. The decreasing diameter and lack of slope produces a deeper water here, and an optimal location for depth measurement for calculation.

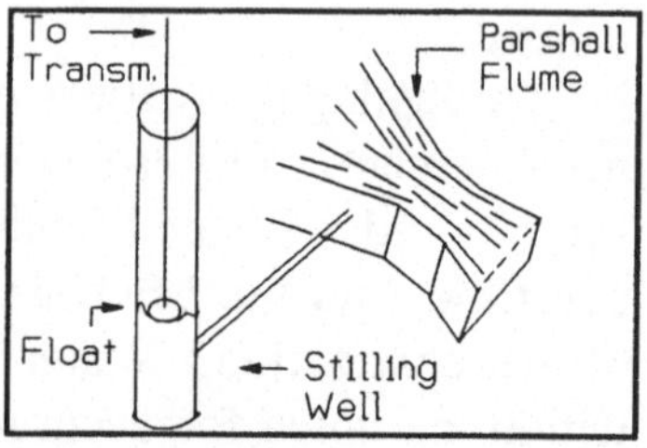

Throat: the narrowest part, with a downward incline. Flumes are sized based on throat width (they range from 1 inch to 50 ft).

Outlet: diverging and sloping upward.

A depth measurement is taken at the first 1/3 of the inlet section with a calibrated stilling well (piezometer at the side of the flume) or one with a float gage, or by overhead ultrasonic unit. The entire depth of the water is measured.

Formulas for flow through Parshall Flumes differ, dependent on throat width. The formula below can be used for widths of 1-8 feet, and applies to a medium range of flows. Maintenance on a Parshall Flume is minimal.

$$Q = 4\ W\ H^{1.52}\ W^{.026}$$

H = depth in stilling well upstream
W = width of throat

trajectory of a jet

Flowmeters are not installed in every location where one might want to measure flow. Fortunately there are some simple methods of estimating quantity. For the locations where water freely exits the end of a pipe or channel, and drops to a surface below, flow can be quickly computed based on measurement of the horizontal and vertical distance of travel after exit.

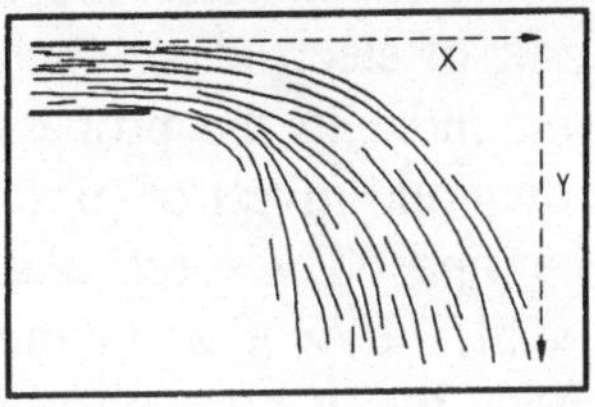

Measure the x and y axis of free fall, and calculate:

$$Q = 4\ A\ \frac{x}{\sqrt{y}}$$

A = the area of flow (this will be the same as the pipe cross sectional area as long as the pipe is full).

These are specialized instances, but more common than one might think. Inplant, for recirculating flows, and for chemical feed, water often flows freely into a tank.

In the absence of flowmeters, discharge can also be estimated by filling or emptying a tank of known volume, and timing the process. For very small flows, clock the time it takes to fill a cylinder or bucket.

pitot gage (pitot tube)

Invented in the 1700's, this is a flowmeter which relies on a direct measurement of Velocity Head, based on differential gage readings, as explained in Chapter 3. The section facing upstream senses Pressure Head and Velocity Head. The downstream pointed side only measures Pressure Head. Subtracting for Velocity Head, this can easily be converted to a flow rate. The Pitot Tube is used to measure system and fire flow, and can be attached to fire hydrants in water distribution systems.

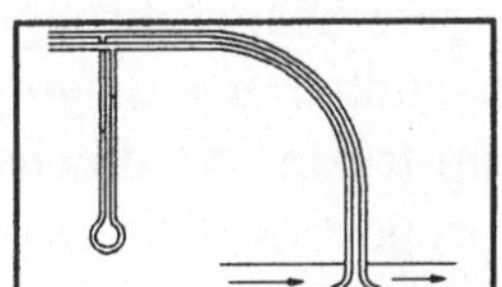

magnetic flow meter

Widely used in wastewater treatment plants, magmeters consist of a set of magnetic coils and an opposed pair of electrodes which are purchased as a flanged pipe unit and mounted in the process pipeline. The magnets are inserted onto crown and invert of pipe, and throw a magnetic field right through the pipe. The electrodes protrude through either side of the pipe wall. It works on the principle that when an electrical conductor (the water) passes through an electromagnetic field, a voltage is induced at right angles to the field, and this voltage is proportional to velocity. It is picked up by the electrodes and then converted to a current signal, which can read out as a flow rate. The magmeter offers no restriction to the flow, can be installed into short runs of pipe and is not affected by liquid properties. The electrodes must be kept free of solids. There is also an open channel version of the magnetic meter, which is used to measure flows in wastewater collection systems.

ultra sonic meters

More recently developed, ultra sonic meters are being used to obtain flow readings in pipe systems, and in open channels.

For closed pipe flow, the transmissive type of meter sends sonic beams through the pipe from opposite transmitter/receivers mounted at a diagonal to the flowstream. The difference between upstream and downstream directed beams is measured, is directly proportional to velocity and is calculated to flow. This meter is installed completely outside the pipe, and the pipe wall thickness must be exactly known. In use for clean water lines only, solids in the water will interfere with the flow reading.

The reflective type of sonic meter is meant for wastewater use. A single sonic beam is sent into the water from a transducer mounted on the pipe; it bounces off solids in the water, and returns at a different frequency because of the water velocity. The magnitude of frequency change is measured, velocity is computed, and then flow.

For open channel flow, an overhead unit sends ultra sonic beams down to the water surface. This is often located at the head of a Parshall Flume in a wastewater treatment plant (where the stilling well would be). The beam bounces off solids in the water and returns to the sensor. Flow is determined based on the time it takes for the beam to return. It is actually measuring the depth of the water and then calculating for flow. Restriction is that there must be solids in the water.

rotameter

Rotameters are adaptable to small flows; they are accurate, but expensive, and are used most often on chemical feed lines. This is a variable area meter, whose operation is based on the continuity principle ($Q=AV$). The flow passes through a transparent, calibrated, diverging tube in which is suspended a float. The velocity of the water carries the float (heavier than water) up the tube until the area is wide enough to slow the velocity to the point where the float remains suspended and stationary. At that point the correct flow is read from the calibrations.

totalizer meters

All the metering devices we have considered to this point have been flow rate meters, meant to provide a reading of cubic feet per second or gallons per minute, for utility process control use. On the other hand, totalizer meters count up the flow as it passes through the unit, for estimating water demand, and for billing.

positive displacement meters

Among the Service Meters (for customer use), positive displacement meters are suitable for residential customers who experience low flows and long periods when no water is used. The operating part is a measuring chamber which encloses a disc or piston. Each time a unit of water passes through the meter, the disc oscillates to pass the flow, and transmits the movement to a register, which records. The register unit is permanently sealed, and contains the register and a reducing gear train. A magnetic contact from the piston to the register unit drives it. Positive displacement meters are accurate at low flows, but have high head loss (5-10 psi), and are too expensive to use for mainline flow. They are normally used in 5/8 inch to 3 inch sizes, and normal flows should not be more than 1/3 of maximum meter capacity; operating them at excessive flows will result in wear and underregistration. Most residential services are metered with 5/8 inch meters that have 3/4 inch connections.

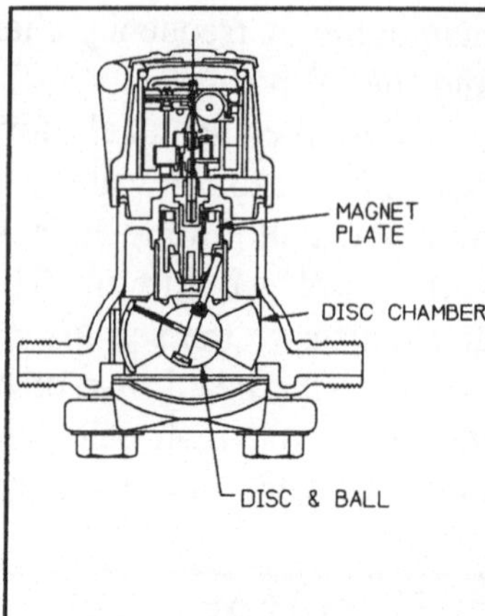

NUTATING-DISK DISPLACEMENT TYPE METER

turbine meters

Turbine meters have a rotor in the measuring chamber which turns in response to the velocity of the water. There is either a direct linkage to the receiver, or the turbine blade may be fitted with a small magnet, and each revolution is picked up by a sensor on the outside of the meter. The movement is registered on a drive train which totalizes flow. Turbine meters are designed to handle large flows. The unit is inaccurate at low flows because small amounts of water slip past the rotor without moving it. Turbine meters generate very little head loss. The newer, more efficient models are called Turbo Meters, or Class I Turbines.

compound meters

Also a customer meter, the compound Meter is most useful where accuracy at both high and low flows is required. Within its body are a turbine meter on the main line, and a positive displacement meter on the bypass. At low flow, water passes through the positive displacement section, and the recording is of this section. The water continues on through the turbine section, but no recording takes place. When the pressure becomes great enough, the bypass shuts off and the turbine takes over, actuating its register. One of the disadvantages of this type of meter is the momentary loss of reading during the changeover from low to high flow. Designing the sections in series, as the one in this diagram, minimizes the loss, for the turbine is already in motion when flow increases enough for it to take over.

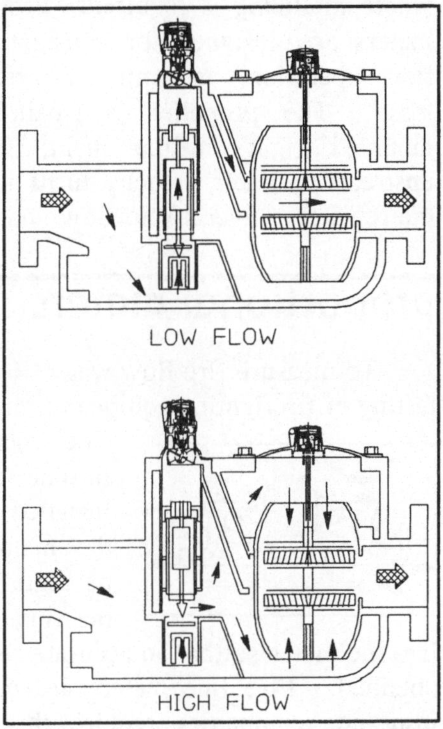

Compound meters are the most expensive of the service meters, but if 10-30% of the flow is on the small side, the savings achieved in recording those low flows will probably be worth the cost. If not, a turbine should be installed.

propeller meters

Similar to the turbine meter, this unit is used in water mains where totalizing is required. Propeller meters are designed for constant high flow operation, and minimal pressure loss. The propeller is small, and mounted right in the main, facing upstream. Water velocity turns it and the register converts it to flow and totalizes.

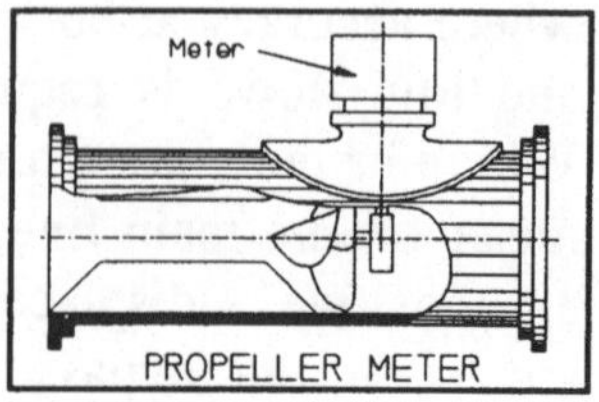

proportional meters

To measure fire flow water, a proportional meter is used. For testing of firefighting equipment, these are installed onto fire hoses. The metering apparatus is a small turbine or positive displacement unit installed into a loop which extends off the main line, and measures a small percentage of the water flowing through the line. A multiplying factor is built into the meter so that an accurate reading of the whole flow can be obtained. Fire line meters are designed this way because fire insurance companies require that no restrictions which could interfere with fire flow be installed onto the line (even if the meter were broken).

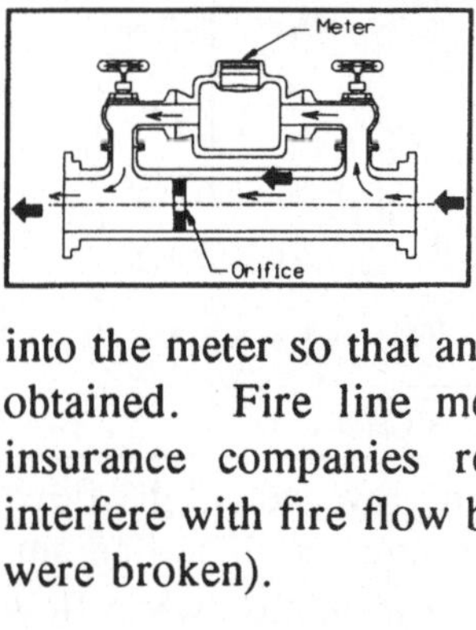

sizing service lines and meters

It is up to the discretion of the water utility to decide which and how many customers will be connected to the community system. Once connected, however, it is the responsibility of the utility to supply adequate pressure and flow to all customers, whether or not it is a metered system. There must be enough water to satisfy peak demand and fire flow, and pressures must be high enough throughout the system to minimize the possibility of backflow. The utility's responsibility extends up to the customer meter. It is not responsible for pressure losses within the customer's building; however, reasonable loss should be allowed for. When installing service to a new development or industry, the utility usually designs the service line so that there will remain in the line a pressure of at least 35 psi immediately downstream of the meter.

The type of demand must be considered. Industry usually requires a continuous flow. Irrigation water is the largest demand, but is of short duration. Domestic demand is the smallest, and use is intermittent.

FIRST: Establish a peak flow demand (gpm needed) for the new establishment. Domestic water demand is due to the use of plumbing fixtures and washdown facilities by the customers. Assuming an average pressure of 35 psi available at the meter, the Plumbing Fixture Value Table lists various fixtures and a flow value for each.

- Take an average dwelling; refer to the table.
- Obtain Fixture Value (flow needed) for each fixture in the dwelling.
- Multiply each by the number of fixtures of that type which will be in the dwelling.
- Add them all up.

This is the potential peak flow demand (gpm). It would be valid if all residents of the dwelling use all these fixtures at once. The probability of this is slim. Therefore, make a correction from the Water-Flow Demand per Fixture Value chart, and obtain a more realistic peak demand value for the dwelling. Multiply this value by the number of dwellings expected of this type.

- Do the same for other types of dwellings in the new area.
- Do it also for the industries.

- Add them all up. This is the actual Peak Demand. Use it as the flow.
- It may be advisable to adjust for projected future demand.

For an industry, the starting point often is the flow that the industry knows it will need; in such cases the entire section above may be eliminated.

NEXT: Size the service line (lateral which runs from the main to the new establishment).

Information Needed:

- Flow capability of main that serves the area.
- Static pressure in main during peak demand periods.
- Elevation difference between main and establishment.
- Length of service pipe from main to meter.
- Friction loss through meter and valves.
- Size of customer's piping and approximate head loss within the building.

To this point we have assumed that we need X gpm, and will want a minimum 35 psi pressure just downstream of the meter.

- Measure elevation difference from main to establishment; convert to psi and add or subtract from main pressure.
- Draw diagram of the new lateral with all fittings and valves. Measure length; choose a diameter. If there is more than one pipe size involved, include all. Draw it out exactly the way it will be installed.
- Convert all fittings and valves to equivalent lengths of straight pipe; add up; add to original pipe length.
- Using the peak demand flow just calculated, and the Hazen-Williams formula, obtain slope, then head loss. Convert to psi.

- Subtract the head loss from the static main pressure. This is the pressure at the end of the lateral, just upstream of the meter.
- If it is an industry, does it require an RP device? Subtract the psi lost from this.

<u>Is there still at least 35 psi left?</u>

If not, choose a larger diameter pipe for the lateral, and try again.

There is one more thing to consider: At this flow and pipe diameter, calculate the velocity. It should be under 15 ft./sec. If it isn't, choose a larger lateral to slow down the water. The necessary flow will still be obtained, the pressure drop will be less, and hammer and corrosion will be minimized.

NEXT: Choose a meter.

Choose a meter of the type and size that is right for your flow and type of use. Meters should not be chosen by pipe size, or by pressure loss. Each meter type and size has been designed for a specific range of flows. AWWA has established standards for cold water meters based on flow capacities. The chart, <u>Meters By Flow</u>, lists various meters, their capacities and uses. The <u>Meter Head Loss</u> listing gives average pressure losses for different meters.

- A meter which is too large for the flow is a waste of money, and will let small flows slip though unrecorded.
- A meter which is too small for the flow it must pass will work too hard and wear out early.
- The meter chosen should have a maximum capacity at least twice as great as the expected flow.
- Check the listed psi loss through the meter which is chosen. Subtract this from the static pressure in the lateral which has just been designed. If there is still 35 psi left, the expected flow will be delivered to the establishment.

Plumbing Fixture Value

Fixture Type	Fixture Value Based on 35 psi at Meter Outlet
Bathtub	8
Bedpan washer	10
Combination sink and tray	3
Dental unit	1
Dental lavatory	2
Drinking fountain (cooler)	1
Drinking fountain (public)	2
Kitchen sink: ½-in. connection	3
3/4-in. connection	7
Lavatory: 3/8-in. connection	2
½-in. connection	4
Laundry tray: ½-in. connection	3
3/4-in. connection	7
Shower head (shower only)	4
Service sink: ½-in. connection	3
3/4-in. connection	7
Urinal: Pedestal flush valve	35
Wall or stall	12
Trough (2-ft. unit)	2
Wash sink (each set of faucets)	4
Water closet: Flush valve	35
Tank type	3
Dishwasher: ½-in. connection	4
3/4-in. connection	10
Washing machine: ½-in. connection	5
3/4-in. connection	12
1-in. connection	25
Hose connections (wash down): ½-in.	6
3/4-in.	10
Hose (50-ft. length - wash down): ½-in.	6
5/8-in.	9
3/4-in.	12

Reprinted from Sizing Water Service Lines and Meters, M22, by permission.

Water Flow Demand per Fixture Value - Low Range

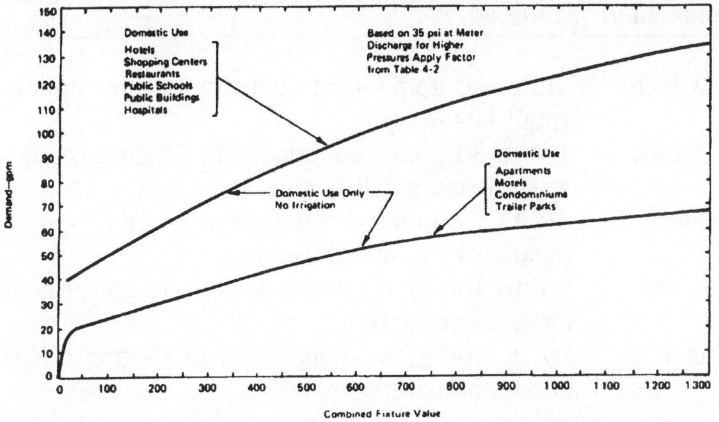

Water Flow Demand per Fixture Value - High Range

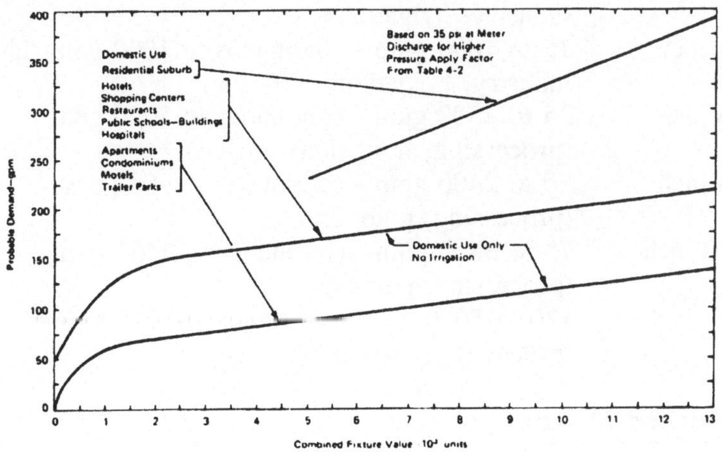

Reprinted from Sizing Water Service Lines and Meters, M22, by permission.
Copyright 1975, American Water Works Association.

Meters By Flow

Displacement Meters

5/8 inch	1/8 to 20 gpm - continuous to 10 gpm (homes, small business)
3/4 inch	1/4 to 30 gpm - continuous to 15 gpm (homes, apartments, small business)
1 inch	3/8 to 50 gpm - continuous to 25 gpm (apartments, small business)
1½ inch	5/8 to 100 gpm - continuous to 50 gpm (motels, large apartments)
2 inch	1¼ to 160 gpm - continuous to 80 gpm (large hotels, housing complex)

Turbine Meters

2 inch	3 to 200 gpm - continuous to 160 gpm (hotels, housing complexes)
3 inch	4 to 450 gpm - continuous to 350 gpm (hotels, industry, irrigation)
4 inch	15 to 1000 gpm - continuous to 1000 gpm (large industry, irrigation)
6 inch	25 to 2500 gpm - continuous to 2000 gpm (processing, irrigation)
8 inch	50 to 2400 gpm - continuous to 900 gpm (processing, pump)
10 inch	75 to 3800 gpm - continuous to 1450 gpm (processing, pumps)
12 inch	120 to 5000 gpm - continuous to 2150 gpm (processing, pumps)

Compound Meters

2 inch	1/4 to 160 gpm - continuous to 160 gpm (motels, schools)
3 inch	1/2 to 350 gpm - continuous to 350 gpm (hotels, large hospitals)

Meters By Flow (continued)

Compound Meters (continued)

4 inch	3/4 to 1000 gpm - continuous to 1000 gpm (apartments, dormitories)
6 inch	1½ to 1000 gpm - continuous to 500 gpm (apartment complexes)
8 inch	2 to 1600 gpm - continuous to 800 gpm (large condominiums)
10 inch	4 to 2300 gpm - continuous to 1150 gpm (large complexes)

Reprinted from Water Audits and Leak Detection, M36, by permission.

Meter Head Loss

Displacement Meters

Size	Pressure Loss (psi at gpm)
5/8 inch	8.0 at 20
3/4 inch	10.0 at 30
1 inch	11.0 at 50
1½ inch	11.5 at 100
2 inch	12.0 at 160

Compound Meters

Size	Pressure Loss (psi at gpm)
2 inch	5.0 at 160
3 inch	4.0 at 320
4 inch	3.0 at 500
6 inch	14.0 at 1000
8 inch	14.5 at 1600

8

centrifugal pumps

Man has always needed to move water from one place to another against the forces of nature. Gravity will move it downhill on a grade. If depth is built up behind, that pressure will move it further. But when pressure is dissipated by friction loss, and when water in the valley is needed on the mountain, the energy to move that water must be artificially created. A pump is needed.

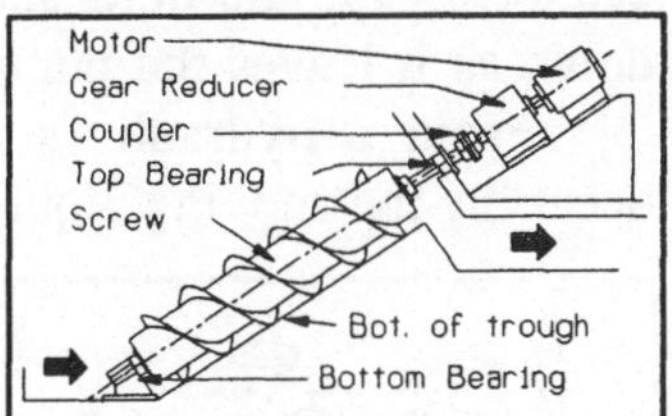

SCREW PUMP

Archimedes invented the screw pump in 287 BC. This physically lifts volumes of liquid, but it has space and height limitations.

It is said that the Roman emperor, Nero, invented the piston pump at about 100 AD. Volume after volume of water is displaced with each stroke, but this is a high energy consumer, and size limits capacity.

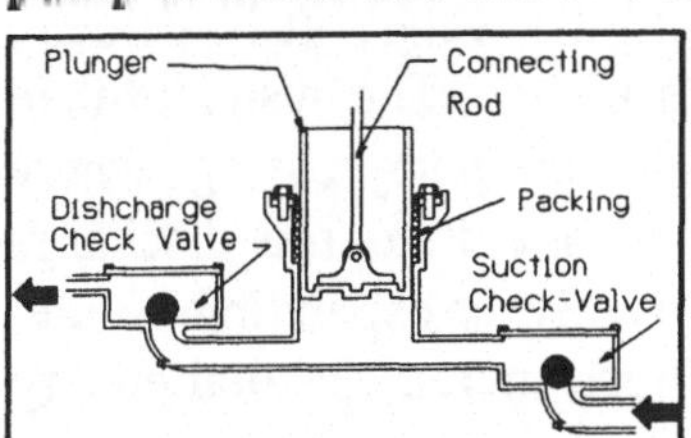

PISTON PUMP

Pumping technology was restricted to these, and variations, until the nineteenth century. The basic concept of the centrifugal pump existed during the 1600's, but development was slow because the high speed drives needed were not available for another 200 years. It wasn't until the 1800's that the first fully functional centrifugal pumps were developed. Since then, we have been able to move great volumes of water with much smaller units than with the pumps previously in use. Centrifugal pumps operate with high efficiency, few parts are required, initial cost is low, and maintenance is relatively easy.

Operation of the centrifugal pump is based on the principle that a high velocity is imparted to the water as it enters the pump, and that velocity is converted to pressure as the water exits the pump. From a mechanical standpoint, water enters the pump through the center, or "eye" of a rapidly rotating impeller. It is moved through the impeller vanes and thrown off their tips with great velocity. Still contained by the pump casing, the water cannot escape the unit, and collects at the volute, or exit, slowing down as it leaves the pump.

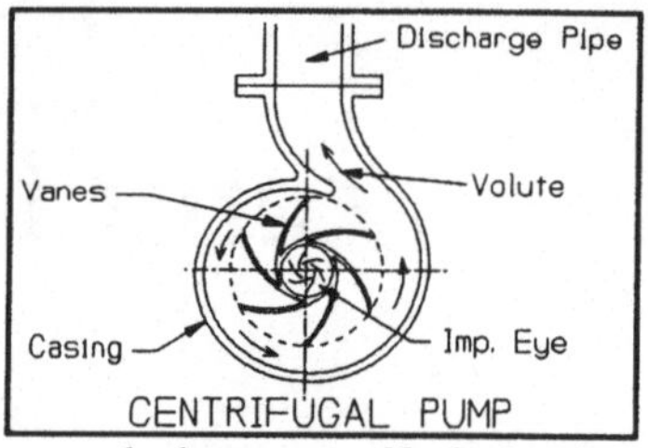

From a hydraulic standpoint, note the energy changes that occur in the moving liquid. As water enters the pump from the suction piping, pressure is low; that's why the pump was installed. Passing from the suction piping through the impeller eye, water velocity increases (smaller area). This actually depresses the pressure even more, and acts to keep water coming in from behind, because of the relatively higher entrance pressure. As the water comes into contact with the rapidly spinning impeller vanes, velocity is increased tremendously. This is the artificial energy which the mechanical action of the pump imparts to the water. It is Velocity Head energy which is increased at this point. The water is then rapidly thrown off the impeller tips against the pump casing, where it begins to slow down, collect, and then exit to the discharge piping through the only outlet, the progressively expanding volute. As Velocity Head decreases, Pressure Head increases; total energy must be maintained, and is merely converted from one form into

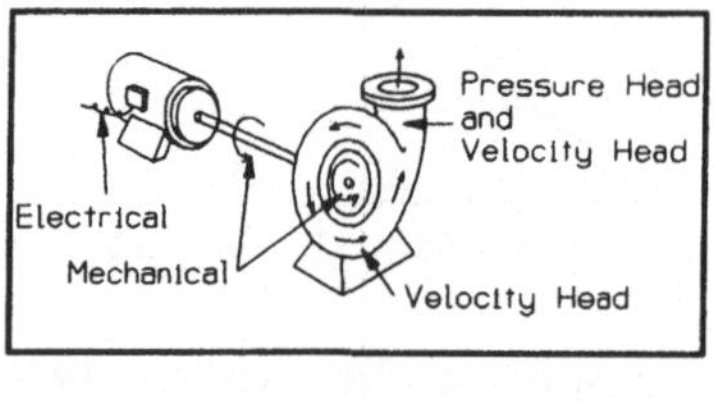

another (from Bernoulli's Theorem). The water now has enough new pressure to lift it the required distance.

The basic centrifugal pump working parts are impeller, casing, and the pump shaft, which connects it to the motor. There are no valves in the operating mechanism, and no close tolerances. Rotating parts do not touch stationary parts, and because of this, there is always some backward slippage of water between impeller and casing, which will vary depending on how hard the pump must work.

> **Note:** A centrifugal pump does not always pump the same amount of water.

The design capacity of a centrifugal pump is determined by the diameter of the impeller eye, and the width of the impeller vanes. The amount of pressure which can be created is controlled by the diameter of the impeller, and the rotating speed (velocity of the impeller rim). The faster it moves, the more water velocity is created, the more pressure will be available as it slows down at the exit.

Pumps with closed impellers are most efficient, and are used for clean water pumping. Pumps with semi-open impellers are used for solids handling water. Pumps with open impellers are sometimes used to pump sludges. Obviously the more open impellers have a better chance of passing solids without clogging.

pump types - position

Vertical Pump

The impeller shaft is in a vertical position. Well pumps are designed like this; impellers are at bottom of the well, and shaft extends upward to ground surface. Submersible pumps, with the motor close coupled to the pump, are vertical pumps used for underwater duty.

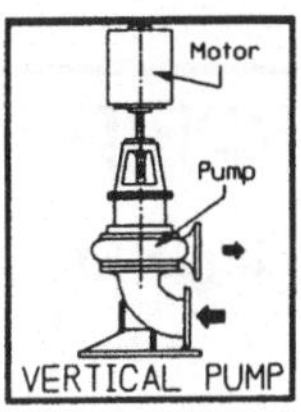

Horizontal Pump

The impeller shaft is in a horizontal position. The typical end suction pump is designed this way.

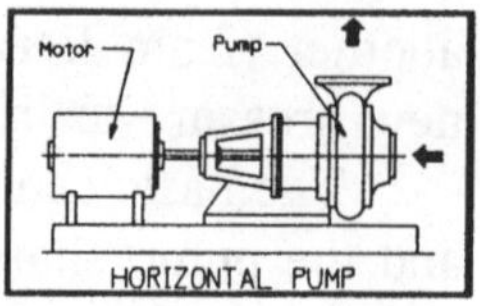

pump types - exit flow direction

Radial Pump

The flow leaves the impeller at a 90 degree angle from the direction it enters the pump. Most centrifugal pumps are radial pumps. The High Service Pump at a potable water treatment plant, which lifts water from the plant to elevated storage is usually a radial pump.

Axial Pump

Water enters and exits the pump on the same plane. Backwash pumps, and Low Service Pumps, which carry water from the source to the treatment plant, are usually axial pumps. They are best suited to pump great volumes of water.

pump types - suction condition

Single Suction

One suction line enters the pump casing.

Double Suction

The suction line splits in two as it enters the pump and water approaches the impeller from both sides. This type minimizes head losses inside the pump; double suction pumps are often seen as distribution booster pumps.

pump types - impellers

Single Stage

One impeller is installed on the pump shaft.

Multi-Stage

These are designed for uses where water must be delivered at high pressures: to feed boilers, membrane processes, and for deep wells. It is designed so that the discharge from one impeller is led through a passage in the casing to the suction of the next impeller.

other important pump components

Stuffing Box and Packing

Inspecting the diagram below, as the shaft extends outside the pump, water is prevented from escaping by the stuffing box, which is filled with rings of packing. The packing acts primarily as a seal, and also as a lubricant for the shaft. It is pressurized by an external source of water at a higher pressure than the water inside the pump. Often this is a small flow of the pump discharge water which is directed back to fill the packing. This water pressure is increased because it has already passed through the pump. Pump sealwater flow is small, but constant, and stuffing box is designed to allow a constant leakage.

Mechanical Seals

This is an alternate to sealwater and packing, and has two components:

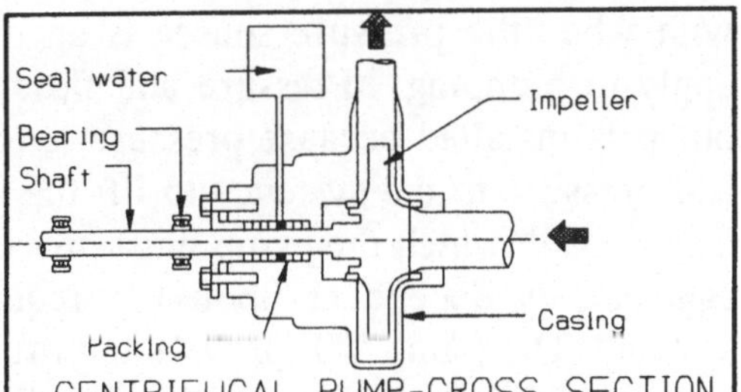

CENTRIFUGAL PUMP-CROSS SECTION

Rotating Assembly: attached to the pump shaft.

Stationary Assembly: positioned permanently on the stuffing box.

The face of the stationary assembly provides a running surface for the face of the rotating assembly. There is no water leakage from a mechanical seal.

Bearings

Bearings take up the stress of the rotating shaft. They hold it steady, and support the impeller. The rotating part of the bearing is locked to the shaft. The housing holds the stationary side of the bearing. There are two types, both of which are in most pumps. However, there is great variation in bearing position, type and number, depending on the pump.

Radial Bearing (line bearing): This one is closest to the pump. It rides free in its own compartment and takes up and down stresses.

Thrust Bearing: this one is farthest from the impeller, closest to the motor. It takes up the major thrust of the shaft, which is opposite from the discharge direction.

pumping characteristics: the pump performance curve

For a given centrifugal pump, the total pressure developed, the horsepower required to drive it, and the resulting efficiency, vary with the discharge. The hydraulic principles which we are familiar with when the pressure source is an elevated tank of water, also apply to pumping. Pressure and flow vary with one another. A pump is installed because pressure is low or nonexistent. It must add pressure to the system - to lift the water and get the job done. This "head" which the pump must provide, or which it must "pump against", is dependent upon the requirements of the particular system. The pump will provide the lift needed, and will overcome the piping headlosses encountered while doing this.

Note: A centrifugal pump will provide enough pressure to overcome lift and losses. No more!

The flexibility is in the flow. If the lift is a great distance, or if piping head losses are great (or both), the pump will do the work required. It will produce enough pressure, but will operate at a much smaller flow than if the lift and losses were smaller. Recall the lack of close tolerances between the impeller and pump casing, which allows backwards slippage of water as the pump operates. This slippage is greater as the head (lift plus losses) increases, and therefore, flow decreases.

Horsepower requirements also vary with flow. Generally, if the flow is greater, the horsepower required to move the water will be greater.

Pump performance is measured in gallons per minute delivered, height to which water is lifted, and efficiency. Each pump manufactured has a designated amount of head and flow at which it will operate most efficiently, and pump purchase should be based on how close to peak efficiency the pump will come in a particular system. This allows the pump to operate with the least strain, both mechanically and hydraulically. A pump which operates well off peak efficiency may do the job, but the result will be excessive energy requirements and shortened pump life.

The interrelations of pump head, flow, efficiency and horsepower are known as the Characteristics of the Pump. These are important elements in pump performance, and they are diagrammed graphically on a Performance Curve. The curve puts these elements together to show a picture of what the pump can do. Each centrifugal pump will be expected to perform as its Characteristic Curve describes. It will not stray from this.

If flow from a pump is measured, head (lift + losses) is calculated, and the characteristics do not match the curve, it is most likely that the losses have been estimated incorrectly.

Inspecting the typical Pump Characteristic Curve, the horizontal scale on the bottom represents the flow (gpm) which this pump will deliver. This is plotted against Head, and creates the H-Q curve, which is the base for most pump curves. The vertical scale on the left represents head, in feet of water. Head is composed of lift plus losses to be overcome by this pump, the total amount of useful energy that will be transferred from the impeller blades to the water. This is Total Dynamic Head, and it registers visibly as the difference between the suction and discharge gage pressures.

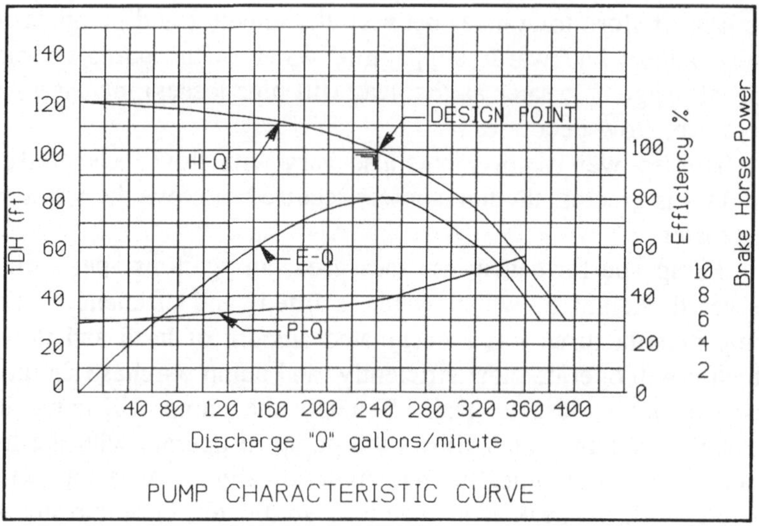

From the horsepower scale on the right is plotted power consumed, also against flow, and makes up the P-Q curve. This is Brake Horsepower, and refers back to Chapter 4, where we first encountered power calculations.:

$$\frac{\text{gpm x hd}}{3960} = \frac{\text{whp}}{\text{pump eff.}} = \text{bhp}$$

The pump curve is actually derived from this formula. Try it. Pick any head on the curve. Read flow, and efficiency at that flow. Calculate brake horsepower with the formula. Then look it up on the curve. It will match.

The Efficiency curve, E-Q, is also plotted against flow. Note that because of internal losses within the pump itself, no pump is 100% efficient. New pumps are usually about 85% efficient, and efficiency decreases with age.

Since all the curves on the graph are plotted against flow, it is easiest when reading these curves, to determine flow first, then proceed up from the flow at the bottom to the curve with the element you desire.

The Rated Head and Flow are those at which maximum efficiency is reached. This is the Design Point, the point at which this pump was meant to operate. If we had purchased this pump for installation into our system, it would be because the head and flow conditions at this pump's design point matched the head and flow demands of our system.

Changing System Conditions

There is a maximum pressure that each pump can achieve. If the pressure needed is greater than that which the pump can deliver, the pump is said to be running at Shutoff Head.

For example, with a pump having the characteristic curve above, and installed into the system in this diagram, if there are no friction losses, the pump will lift the water 120 ft. up the pipe, since 120 ft. is shutoff head for this pump. The pump will keep running, but the water level will remain there. The pressure of the discharge water will be equivalent to the maximum pressure that this pump can achieve, and so there will be no flow. The water inside the pump will just circulate around and around, heating up, and will result in damage to the pump. This is the same as operating the pump with the discharge valve closed. Since the water isn't going anywhere, horsepower requirements are lowest at this time.

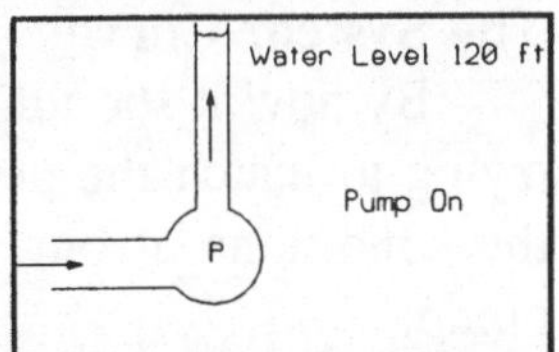

Pump Operating At Shutoff Head

It is often desired to start up a centrifugal pump in this condition, with the discharge valve closed, in order to start the flow moving slowly, then gradually opening the valve. This minimizes surge upon startup and gives the water in the suction line a chance to overcome inertia and get moving into the pump.

Inspecting the characteristic curve again, try extending the H-Q curve down farther than it is drawn on the right, and you can imagine the pump running at high flow and negative head. This can happen upon starting a pump. A huge initial flow cannot overcome the inertia of the water in the suction pipe, and the water will split, or pull away from itself, creating a vacuum and drawing air into the pump. Centrifugal pumps have great difficulty expelling air, which may cause damage to the pump. It is better to start up with the discharge valve closed, or have the valve automatically operated so that it opens slowly as the pump starts. This phenomenon of high flow/negative pressure should not be new to us. The same thing occurs in a water line when flow suddenly increases, upon a main break, or a hydrant fully and suddenly opened: large flow at once, vacuum created upstream, water separates. This invites a chance for backflow of cross connected contaminant into the system.

When a pump is installed to discharge water horizontally through a pipe, there is no lift required, and the pumping head will be equivalent to the head losses only.

The System Curve

By now it should be apparent that in choosing a pump, we are trying to match the design point of a particular pump's curve with the conditions of our system, and if it does match, purchase that pump.

The system is the arrangement of pipe, fittings, and equipment through which the water will flow. The easiest way to match H-Q characteristics of the system with H-Q characteristics of various pump curves, is to develop a System Curve. The system head consists of lift plus losses. Lift is set, and can be measured, but head losses depend upon the amount of flow moving through the pipe. How can the effect that the system head loss will have on a prospective pump's flow be determined, when the head loss in the system cannot be calculated until after the flow is known?

A system curve must be developed, with calculated heads based on a range of flows; then the characteristics of the system will be known, the actual conditions under which a pump will operate. The H-Q point at which it is desired to operate the pump can be marked. Several possible pump curves (representing several different pumps), can be inspected and superimposed onto the system curve, and the one that matches the desired point, at a place on the pump H-Q curve which is at or near its design point - that is the pump which should be purchased.

To develop the system curve, choose a flow lower than that which it is desired to operate at, and calculate the head losses at that flow (Hazen-Williams); add this to the required lift, and mark it as "head" on the graph. Do this with several different increasing flows, and draw the graph from point to point. This curve now represents the characteristics of your system. Note that system curves do not usually start at the bottom of the graph. The curve itself represents only the head losses. The level it starts at represents the lift.

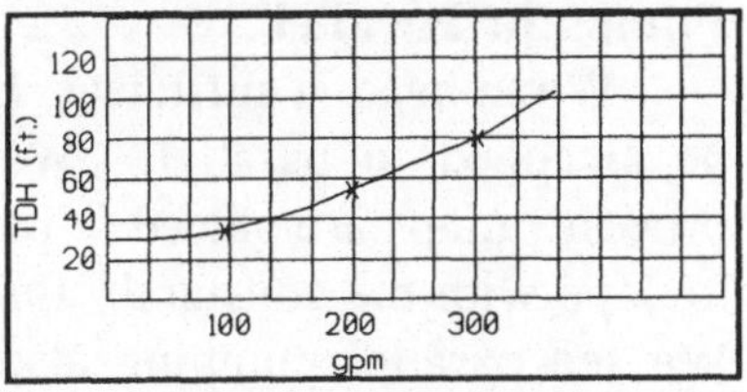

Pumps in Series

When system heads are too great for one pump to overcome, pumps may be placed in series, one after another in the same pipeline. This puts more than one source of pressure into the system, and it increases the head. However, these pumps are pumping the same water through the same system. Series pumping is not meant to increase the flow.

We are familiar with:

Booster pumping: the system has expanded; pipelines have been extended, or an apartment building has been added, and now more pressure is desired to lift the water that extra distance. A booster pump is installed at a low pressure area to raise the head to a safe and practical level for consumer use.

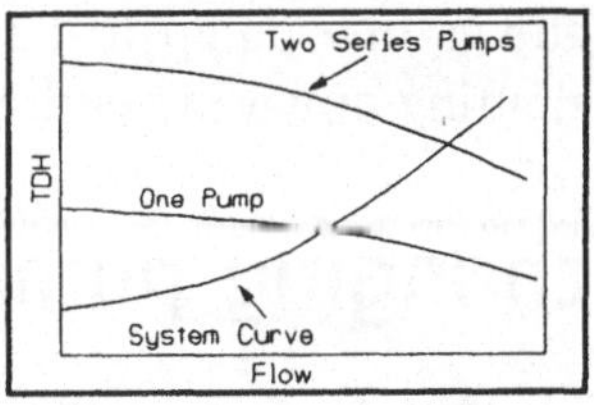

Multi-Stage pumps: a case of impellers in series in the same pump casing, hydraulically a series pumping setup. High pressure pumps, and deep well pumps must achieve great heads. The discharge of one impeller is directed into the eye of the next impeller. The flow is constant throughout, but at each stage the head increases.

Pumps in Parallel

When head is sufficient, but more flow is needed, pumps may be arranged in parallel. Wastewater lift stations and treatment process pumps are placed in parallel in order to cope with the constantly fluctuating flow. We see parallel pumping also in industrial processing, where there exist specific needs for volumes of water for washing, rinsing, processing a product.

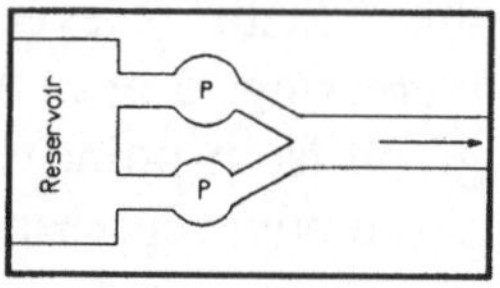

From the standpoint of each pump when two pumps are working in parallel, the system requirements are the same, and must be met. Each pump achieves the same head for its flow, but since there are two of them, the system is getting twice as much water.

Pumps arranged in parallel increase the flow, but the head remains that of one pump working. When combining pump curves to determine characteristics, the combined discharge is the sum of the discharges of both pumps at that head.

Parallel or series operations allow the operator to be flexible enough in pumping capacities and heads to meet requirements of system changes and extensions. Multiple pump operation also allows one pump to be taken out of operation for repair without shutting down the entire system.

changing pump performance

Up to this point system changes and their effect on pump operation have been considered. However, the pump itself has some flexibility for change. Pump characteristics can often be adjusted to meet the system needs, and yet maintain peak efficiency. Each pump characteristic curve applies to that pump with the provided impeller, operating at the stated speed. Impeller size and pump speed are printed on the nameplate, but in many pumps these can be changed. Changing rim speed of the impeller may be the most economical way to make adjustments; it will predictably affect head, flow, and brake horsepower. This actually changes the rating of the pump and develops an entire new pump characteristic curve for it.

Pump speed can be changed in either of two ways:

Change pumping RPM: with a fixed gear drive, adjusting shaft speed up or down. Extra space is required for this equipment.

Change impeller size: many pumps can be fitted with a number of different sized impellers. With a smaller impeller, for instance, the pump shaft is turning at the same rate, but the velocity at the rim of the impeller decreases. Not all pumps are originally purchased with the largest sized impeller installed. Many start with a smaller size, which better matches system characteristics at the time and a larger impeller is installed later, to adjust to system changes.

variable speed pumps

For a system whose needs change frequently with operation a variable speed pump may be the best choice; the speed of the pump will change to match the system needs. For instance, if the system must alternately deliver flow at different capacities, or if static lift is small, and head losses are great and constantly changing, the pump head of a variable speed pump will vary with system head, and the pump chosen will operate efficiently at several different capacities.

Advantages of using variable speed pumps are in the energy savings obtained by operating close to peak efficiency regardless of changing system needs, and in the ability of the pump to discharge less water when demand is for less, with resultant smaller horsepower demand. These pumps are used for wastewater pumping, especially at lift stations, where flows vary widely. They are operated in response to information received from tank level controls; pump speed will be what is needed to keep the proper pressure differential. Disadvantages of variable speed pumps are high cost and greater maintenance.

priming a pump

A major deficiency of the centrifugal pump is that the pump chamber must be filled completely with water upon startup in order for it to function correctly. If there exists a suction head (positive pressure on the suction side of the pump) the unit will always remain full, whether on or off, but with a suction lift, water tends to run back out of the pump and down the suction line when the pump stops. If the casing is filled with air or vapor, the impeller cannot create enough vacuum upon starting to draw water back into the unit, and the gas will just circulate around in the pump. The heat produced by the pump's mechanical action has no flowing water to dissipate it, and both pump and motor will overheat in a short time. It is always best to provide a positive suction head for a centrifugal pump, but if that is not possible, the pump must be separately primed (filled with water) each time it is started.

Several different types of priming devices have been developed:

Foot Valve: a specialized check valve at the end of the suction line. It rides open when the pump is operating but shuts as the pump stops, trapping water in the unit and piping. Recorded efficiency varies; the valve must seat correctly.

Vacuum Pumps and Ejectors: separate units attached to the main pump casing, they create the vacuum needed to fill the pump before startup. These are seen installed with low lift pumps at water treatment plants; the disadvantage is extra capital cost and maintenance.

Priming Chamber: a permanently fitted reservoir of liquid above the pump, it is usually filled by a small recirculated flow of pump discharge water, and its contents empty into the pump chamber as the pump starts. This is a bulky unit; it must be able to fill the pump and associated piping completely.

cavitation

Water naturally contains a small gaseous phase, present as tiny bubbles or vapor molecules, which are in constant motion and always trying to escape the fluid. The percentage of this gaseous phase which remains in the water without evaporating, depends on the water's tendency to evaporate. At normal temperature in an open vessel, atmospheric pressure upon the water minimizes this tendency; the vapor areas are fairly stable, and the water remains in the liquid phase. At elevated temperatures, or under conditions of lowered air pressure, the water's tendency toward vaporization increases. At higher elevations water boils at lower temperatures.

As water enters a pipe and pushes the air out, this evaporative tendency of the liquid becomes by comparison more significant. If the water is under water pressure, instead of air pressure, the vaporizing tendency is minimal, but only because normal inpipe water pressures are usually a good deal more than atmospheric pressure (at 50 psi, water will not boil until a temperature of 300 degrees F is reached).

This tendency to vaporize, or the pressure that the water exerts on its surroundings, is called Vapor Pressure. At 60 degrees F, water has a vapor pressure of .256 psia - very small. At 212 degrees F, water has a vapor pressure of 14.7 psia, the same as the atmospheric pressure upon it, and it boils.

When the pressure of the atmosphere falls below the vapor pressure of the water, the amount of vapor phase rapidly increases, bubbles are formed, and the water begins to flash into vapor. These pressure changes are usually momentary; as pressure equalizes in the system and returns to the area, the low pressure condition is eliminated. The vapor bubbles collapse, or implode, quite forcibly, against the surface of the surrounding metal, chipping and disintegrating that metal, bit by bit. This phenomenon is referred to as Cavitation.

The effects of cavitation are most damaging in areas where it is occurring repeatedly. In pumping operations employing a suction lift, there is no positive depth of water to provide an initial pressure to the water entering the pump, and it is here that cavitation is most significant. It is only atmospheric pressure on the intake reservoir (14.7 psia, or 34 ft.) which pushes the water in to fill the suction pipe.

Theoretically, the water level on the suction side could be as much as 34 feet below the pump, and water would still come up the suction pipe to the pump. This amount of absolute pressure needed to get the water into the pump is called Net Positive Suction Head (NPSH). It is the total amount of energy, in Absolute feet of water, at the pump centerline.

If this Net Positive Suction Head should drop below the vapor pressure of the water entering the pump, the water will start to flash into vapor at the pump entrance, eye, and impeller vanes. Bubbles will form and then be forcibly collapsed as the water pressure suddenly increases near the pump exit. Cavitation will occur, with resulting chipping and destruction of internal pump parts. Operated this way for a length of time, the vibration caused will also damage bearings, shaft and seals.

How does this theoretical 34 ft. of water pressure (14.7 psia) become less than the vapor pressure of the water? It does happen because these are both only theoretical, and static quantities.

There are friction losses within the mechanism of the pump itself, and manufacturer's data will supply a specific amount of minimum positive Absolute pressure which must be present at pump entrance in order to overcome these head losses, and keep the pump running efficiently, and without cavitation. This is the Net Positive Suction Head Required (NPSHR). It is a characteristic of each pump and is often inserted as a NPSHR curve on the pump performance curve. It is the amount of Absolute pressure needed to use this pump.

When considering pump purchase, it must be then determined how much pressure the system will provide. This amount is the

Net Positive Suction Head Available (NPSHA), a characteristic of the system. It is defined as the energy actually available, in feet of Absolute pressure, at the inlet of the pump.

NPSHA is calculated for each installation with a simple formula:

NPSHA = (Atmos. Press. + Velocity Hd.) - (Lift + Losses + Vap. Press.)
positive pressure forces — negative pressure forces

> **Note:** All of the elements of the formula are recorded in Absolute Feet of Water.

Atmospheric Pressure: this may be 14.7 psia at sea level, but it is substantially lower in high altitude areas.

Velocity Head: when operating with a large suction lift, Velocity Head may be significant enough to count. Use the water velocity at the impeller eye; this is the area of lowest pressure and highest velocity, and usually where cavitation starts.

Suction Lift: varies with the installation, and may also vary with drawdown if water is taken from a tank. Use the lowest level.

Suction Losses: these depend upon the flow, and will vary with it as the pump operates. Use the maximum flow the pump will be discharging at any time.

Vapor Pressure: varies with the temperature. For cold water flow efficiently pumped, .59 ft. can be used. Hot climates, overheated pumphouses, direct sunlight all contribute to increased vapor pressure.

A cavitating pump usually makes a noise like rolling marbles, gravel grinding, or balloons popping. When excessive, it can be very loud and rasping. Sometimes it can occur with no noise at all, and perhaps is then most serious, for it goes undetected.

There are other factors which may contribute to cavitation within a pumping system:

Entrained Air

The velocity of the water moving through a pumping system does create a slight educator effect, and can easily suck outside air into the system on the suction side through pump packing, suction valve packing, flanged joints and shaft sleeves. Cavitation caused by the presence of entrained air cannot be calculated, and is sometimes more difficult to detect than vapor caused cavitation. Even in cases where low pressure conditions are not severe, air pockets formed inside a pump can become large enough to cause vapor lock, or air blockage. The air cannot be expelled by the pump and capacity will be greatly reduced.

Insufficient Submersion

If the suction line is not submerged adequately in the suction tank, vortexing may be set up around the pipe, drawing air right in with the water.

Shallow Suction

A suction tank filling rapidly as the pump removes water, especially if it is waterfalling in, will trap excessive air, and it may short circuit right into the pump.

Slope of Associated Piping

If the discharge line is lower than the pump centerline, and the discharge valve is opened before the pump is started, vapor can escape backwards into the pump. The same can happen if the suction pipe is not designed with an inclined slope. Air pockets will form in the line, and as they build up, now and then will flush into the pump.

Reducers: At pump entrance the suction pipe decreases size through a cone shaped fitting called a reducer, and at discharge increases size again by the same method. Concentric reducers are only used on the discharge side. Eccentric reducers should always be used at suction entrances. When the supply line enters from below, an eccentric reducer with flat side up should be used at suction. It leaves no chance for air to collect in the pipe at pump entrance. When the supply line enters from above (positive suction head), an eccentric reducer with curved side up should be used at suction. This gives any gas collected there a chance to slide back up the suction line away from the pump.

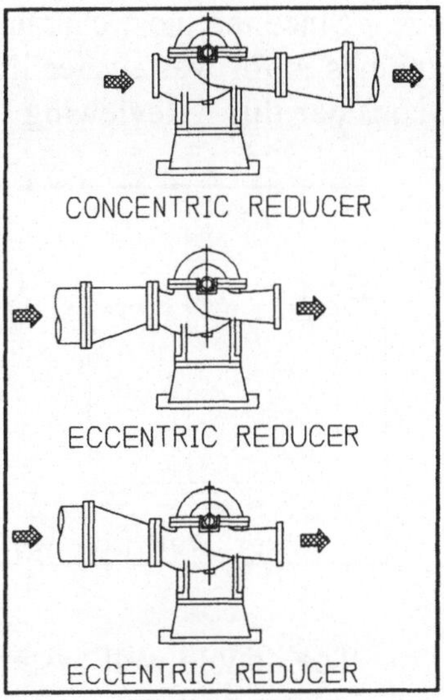

The above sections have referred to water vapor formations from within the liquid, and entrained air entering from the surroundings, but the same principles, and damage, can be applied to any gas which becomes trapped in pump or associated piping. In wastewater pumping it is likely that gases other than water vapor or air will be formed within the wastewater, and lead to cavitation. Wastewater pumps should always have a flooded suction, if possible.

specialized pumping systems

Submersible Pumps

Submersible pumps see increased service each year for many applications. They are now commonplace for sewage and stormwater pumping; submersibles remove urban drainage water to and from retention basins, are used to irrigate farmland and golf courses, and to replenish wetlands. They provide good low pressure pumping for transfer applications: pond to pond, from ditches, and for construction dewatering. In potable water service, submersible pumps provide high pressure for deep well pumping when equipped with multi-stage impellers.

Since the most characteristic and practical use for submersible pumps is for wastewater lift station pumping, we should especially consider this. Reviewing the two basic types of lift stations:

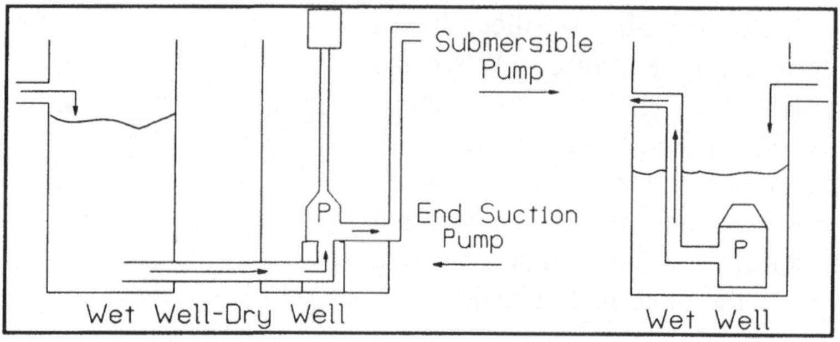

The double well system employs an end suction pump, which takes water with a positive head from the separate wet well, and pumps it through a force main to the surface. The dry well section of the installation houses pump, motor, control panel, and all associated equipment. This structure often extends well above ground as the pumphouse.

The single well system has a wet well only, with a submersible pump installed at the bottom. The pump is submerged, but control panel, including starters and alarm system for high water level, is above ground.

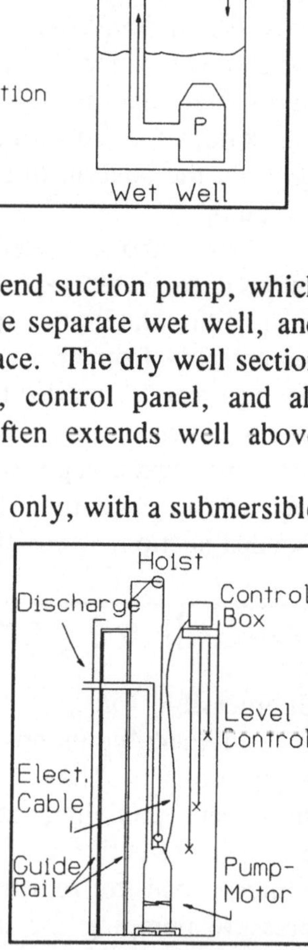

Submersible pumps are vertical, close-coupled heavy duty centrifugal pumps designed to work submerged in the water they are pumping. There is no suction piping, and the pump impeller is at the very bottom of the unit. Both motor and pump are submerged. The motor is watertight, and double sealed from the surrounding water by oil or air. Wastewater submersibles are designed to be non-clogging, and some are specially designed as grinder pumps.

There are many advantages to using a submersible pump. With no pumphouse needed, space and equipment savings are significant, along with aesthetic improvement. The control panel at ground level can be locked and/or sheltered, or installed at a remote location. Submersible pumps are quiet; they operate below ground and underwater. There is no danger of the motor flooding. There is no suction lift with its risk of cavitation. The surrounding water constantly cools both pump and motor, making them quite efficient.

Deep Well Pumps

The typical deep well pump is a vertical turbine pump. This is a centrifugal pump resting on a solid foundation with a shaft extending all the way down the well. Multi-stage impellers are arranged near the bottom of the shaft. As with submersible pumps, vertical turbine pumps have no suction piping; the well water level must be over the lowest pump bowl. Each impeller stage increases head as the water moves upward. The well must be straight and vertical, and correct alignment of shaft sections and column pipe is important.

Submersible pumps with multi-stage impellers are also widely in use as deep well pumps, and are ideally suited to crooked wells.

Jet Pumps

These are inexpensive centrifugal pumps which provide extra suction lift for the operation of small wells which must draw water up from greater than 20 ft. below the pump. Installed on the suction end at the bottom of the well is a venturi apparatus which increases the Velocity Head of the water enough to get the water up the extra distance to the pump.

pump appurtenances: the pressure gage

A pressure gage on a pump should be installed on the discharge piping as far from the pump as practical. If mounted on the pump casing, it will not read accurately, as the energy in the liquid has not yet been fully converted to pressure. A properly operating gage can be a handy tool to diagnose pump problems. If readings begin to show a steady increase in pressure, there may be an obstruction building downstream from the gage. If gage readings show a steady decrease in pressure, check the possibility of a leak or break in the pipe downstream. If this turns out to be the case, turn off the pump and read the gage pressure. The reading will show the elevation of the break.

A discharge gage can be used to check the actual shutoff head. Close the discharge valve, and leave the pump running. The gage will read the true shutoff head. If this value is lower than the shutoff head on the pump curve, there may be internal pump wear or damage.

A discharge gage will yield the approximate flow rate. Read the gage, and refer to pump curve for the flow. (If there is suction head, subtract from gage reading; if there is suction lift, add to gage reading).

pump operation guidelines

Selection:

In selection of any pump, know the exact nature of the liquid which will be pumped (specific gravity, vapor pressure, entrained air, temperature, corrosive properties, type and concentration of solids). In wastewater applications, solids are most important; at a point they change the specific gravity of the water, and it will require more horsepower to do the same amount of work. For abrasive solids, special pump linings may be needed. Stringy materials may require special impellers. Advise the manufacturer of the water characteristics, and the capacity needed.

Analyze suction and discharge conditions. Know whether service will be continuous or intermittent, what type of drive will be used, and what power supply is available.

Be aware of location conditions (space, elevation, population density), capital and operating costs, and transportation limitation.

Select the pump which will yield the required capacity under the worst operating conditions.

Installation:

Install the pump in a light, dry, clean location, and as low as possible. The foundation should be rigid, the driver properly aligned. Appurtenances should include vent valves at high points on the line, drain valves, relief valves, bypass connections, gages, flowmeter, and a source of seal water. Suction piping should be of large diameter and minimum length, and should always be larger than inlet port of pump. Make sure parts and standby equipment will be available.

Operation:

Follow instruction manual on startup and shutdown. Operate the equipment within the proper range of flows, pressures, temperatures. Do not throttle suction to reduce pump capacity. Do not use excessive lubricant or cooling water. Do not run it if it is noisy. Avoid sudden temperature changes. Run spare equipment occasionally. Run the pump at highest efficiency possible to prevent bearing failure. A pump handles liquids. Keep air out!

Maintenance:

Perform scheduled maintenance and inspections. Do not open equipment unless diagnosis indicates need. When this is necessary follow manufacturer's instructions. Examine for corrosion and wear, clean surfaces, restore packing to proper condition (the weakest part of the pump is the stuffing box). Be careful mounting bearings. Keep a complete record of inspection and repairs.

exercises

chapter 1

1. The water in a tank weighs 820 pounds. How many gallons of water are in this tank?

   Ans. 98.3 gallons

2. An alum solution (density = 83 lb./cu.ft.) is pumped into a chemical treatment system at a rate of 3 gpm.
   a. How many pounds of alum are fed in one day?
   b. What is the Specific Gravity of the alum solution?

   Ans. a. 47936 lb./day
   b. 1.33

3. A digested sludge whose Specific Gravity is 1.25 is pumped at 50 gpm to the drying beds.
   a. How many pounds of digested sludge are applied to a bed which is 30 ft. wide, 60 ft. long, and 10 inches deep?
   b. How long does it take to fill this bed?

   Ans. a. 116532 lb.
   b. 3.7 hrs.

4. A heat exchanger is being constructed to maintain temperature in the anaerobic digester. Sludge will pass through a coiled 4 inch diameter pipe 160 ft. long which is set inside a hot water

tank 4 ft. by 4 ft. by 4 ft. high. How many gallons of water will be needed to fill this tank?

Ans. 376 gallons

5. A core sample is taken from the sand filter to check for percent mudball accumulation. The sampler is 6 inches in diameter and 12 inches long. After sifting out the sand, the mudballs are immersed in a 2 inch diameter graduate holding one liter of water. If the water level in the graduate rises four inches, what is the percent mudballs in the filter?

   Ans. 3.7%

6. A flow of 1 MGD occurs in an 8 inch diameter pipeline. What is the water velocity?

   Ans. 4.4 ft./sec.

7. What diameter pipe is required to carry a flow of 40 MGD if the water velocity is 4 ft./sec.?

   Ans. 4.5 ft. (54 inches)

8. The velocity of flow in a 12 inch diameter pipe is 100 ft./min. If it discharges through a nozzle three inches in diameter, what is the velocity of flow in the nozzle?

   Ans. 26.7 ft./sec.

9. What is the gpm flow in a channel 4 ft. wide and 2 ft. deep if the water travels at a speed of 3 ft./sec.?

   Ans. 10771.2 gpm

10. A 6 inch diameter pipe carries 2.87 cfs. The pipe branches into a 2 inch diameter pipe and a 4 inch diameter pipe. If the velocity in the 2 inch diameter pipe is 40 ft./sec., what is the velocity in the 4 inch diameter pipe?

    Ans. 23.3 ft./sec.

chapter 2

1. A gage at the bottom of a standpipe reads 105 psi. How deep is the water in the standpipe?

   Ans. 242.6 ft.

2. A 100 ft. diameter cylindrical tank contains 1.5 MG water.
   a. What is the water depth?
   b. What psi pressure does a gage at the bottom read?

   Ans. a. 25.6 ft.
   b. 11.1 psi

3. The pressure in a pipe is 65 psi.
   a. What is the pressure in feet of water?
   b. What is the pressure in psf?

   Ans. a. 150.2 ft.
   b. 9361 psf

4. What is the head on a pipe whose pressure reads 6234 psf?

   Ans. 100 ft.

5. A gage on the suction side of a pump shows a vacuum of 10 inches of mercury.
   a. What is this pressure in feet of water?
   b. What is this pressure in psi?

   Ans. a. -11.3 ft.
   b. -4.9 psi

6. A pressure gage positioned 7.5 ft. above a pipeline reads 200 psi. What is the psi pressure in the pipe?

   Ans. 203.2 psi

7. What must the pressure in a water pipe 30 ft. below a closed faucet be to give a pressure of 35 psi at the faucet?

   Ans. 48 psi

8. A tank full of water is 15 ft. long, 10 ft. wide, and 12 ft. high.
   a. What will a pressure gage at the bottom read (psi)?
   b. What is the force of water on the bottom?

   Ans. a. 5.2 psi
   b. 112320 lb.

9. A water storage tank on top of a building with a 15 ft. water depth supplies the cylindrical hot water heater in the basement, 50 ft. below. The heater is 6 ft. tall and has a diameter of 4 ft. What is the average force on the heater walls?

   Ans. 290,984 lb.

10. A tank is 10 ft. by 15 ft. by 25 ft. water depth. What is the pressure at the bottom:

a. in psig	**Ans. a. 10.8 psig**
b. in psia	**b. 25.5 psia**
c. in psfg	**c. 1558 psfg**
d. in psfa	**d. 3672 psfa**

chapter 3

1. A flow of 2000 gpm takes place in a 12 inch diameter pipe. Calculate the Velocity Head.

   Ans. .5 ft.

2. If water is flowing at a velocity of 3.5 ft./sec. in a horizontal line at ground level, from a 250 ft. elevated tank which holds water 20 ft. deep, what are the:

a. Pressure Head	**Ans. a. 20 ft.**
b. Velocity Head	**b. .19 ft.**
c. Elevation Head	**c. 250 ft.**
d. Total Head	**d. 270.19 ft.**

3. What is the Velocity Head in a 6 inch diameter pipe connected to a 1 ft. diameter pipe, if the flow in the larger pipe is 1.42 cfs?

   Ans. .81 ft.

4. An industrial water treatment process employs upflow pressure filters, which have a media depth of 4 ft. The pressure loss created by turbulence and friction in passing through the media is 8 ft. If the gage reading on the 6 inch diameter influent line is 65 psi, what is the gage reading on the 6 inch diameter effluent line?

   Ans. 60 psi

5. Water is carried from a tank holding a 25 ft. depth through a 12 inch diameter pipe. Flow is free at 25 cfs at the end of the pipe. What is the head loss due to friction?

   Ans. 9.2 ft.

6. A 6 inch diameter pipe is slanted upward and discharges water into the atmosphere. The flow rate is 1.963 cfs and the pressure at a point in the pipe 6 ft. below the discharge end is 33.5 psi. Calculate the total head loss between that point and the discharge end.

   Ans. 71.4 ft.

7. If the water pressure lost in traveling from the storage tank, which stands 50 ft. off the ground, to the new housing development is 43.3 psi, how many feet of water in the tank should be provided in order to service the housing development with a minimum of 35 psi pressure?

   Ans. 119.3 ft.

8. A high service pump will lift water 90 ft. into an elevated tank. The gage on the pump discharge line reads 45 psi. What is the head loss due to friction?

   Ans. 14 ft.

9. A 4 inch diameter pipe directs a jet of water vertically up into the air. If the rate of flow is 2000 gpm, how high will it go?
   Ans. 42.3 ft.

10. A fire pump delivers water through a 6 inch main to a hydrant. A 3 inch diameter hose is connected to the hydrant. It terminates in a 1 inch diameter nozzle. The nozzle is 10 ft. above the hydrant and 60 ft. above the pump. If friction losses are 28 ft. from pump to base of nozzle, what gage pressure at the pump discharge is necessary to throw a stream of water 80 ft. vertically above the nozzle?
    Ans. 73 psi

---

chapter 4

1. Water is pumped at 500 gpm to a water storage tank on the roof of an industry. The gage on the pump discharge line reads 95 psi. The difference in elevation between the gage reading and the water level in the tank is 90 ft. What is the friction loss (ft.) in the piping?
   Ans. 130 ft.

2. A wet well pump operating with a suction lift has a Total Dynamic Head of 150 ft. If the Dynamic Discharge Head is 130 ft. and the friction head loss on the suction side is 10 ft., what is the vertical distance between the water level on the suction side and the centerline of the pump?
   Ans. 10 ft.

3. A flow of 1840 gpm is to be pumped against a head of 40 ft. What is the water horsepower required?
   Ans. 18.6 hp

4. A 12 hp pump delivers 1200 gpm. Assuming the pump is 100% efficient, what is the pressure against which the pump is operating?
   Ans. 17.2 psi

5. A pump (75% efficient, motor is 82% efficient) delivers 825 gpm against a Total Dynamic Head of 95 ft. Find the:
   a. Water Horsepower — **Ans. a. 19.8 hp**
   b. Brake Horsepower — **b. 26.4 hp**
   c. Motor Horsepower — **c. 32.2 hp**

6. A motor draws 2238 Watts and costs $1.65/week to operate the pump. If the rate per Kilowatt hour is $.015, how many hours per week is the pump in operation?
   Ans. 49 hr./wk.

7. A pump is discharging 700 gpm against a head of 82 ft. The wire to water efficiency is 83% and the power rate is $.08/Kw.hr. What is the power cost for 14 hours of pumping?
   Ans. $14.59

8. A water level gage on a 500 gpm well pump reads 52 psi when the pump is off. When the pump is on, the gage reads 39 psi.
   a. What is the Drawdown?
   b. Calculate Specific Capacity
   Ans. a. 30 ft.
   b. 16.7

9. A tank 20 ft. in diameter and 10 ft deep is half full. Flow entering the tank is 100 gpm; 50 gpm is being pumped out. How long does it take to fill the tank?
   Ans. 235 min.

10. A 75,000 gallon elevated storage tank is one third full. The wells are pumping at 500 gpm. The distribution demand is 200 gpm. How long before the tank is full?
    Ans. 2.8 hrs.

chapter 5

1. A 36 inch diameter pipe carries a flow of 800 gpm. If C=100, find the slope (ft./ft.).

   Ans. .00001 ft/ft.

2. A flow of 10 cfs occurs in a rough concrete pipe of 30 inch diameter. What is the friction head loss in a 2000 ft. length of the pipe?

   Ans. 1.2 ft.

3. What minimum sized lined steel pipe should be used to carry a flow of at least 10 cfs if the friction loss is to be 25 ft. over each mile of pipe length?

   Ans. 18 inch diam.

4. In a test of 20 inch diameter cast iron pipe, the flow is steady at 2.2 cfs, and the Hydraulic Grade Line dropped 4 ft. in a length of 8000 ft. of pipe. What is the value of C?

   Ans. 80

5. Determine the discharge through a new cement lined steel pipe of 1 ft. diameter if the pressure drop is 2 psi per 1000 ft. of pipe.

   Ans. 3.33 cfs

6. One thousand feet of 8 inch diameter pipe is extended by 1200 ft. of 10 inch diameter pipe. The velocity through the 10 inch pipe is 2 ft./sec. If C=100 for both pipes, what is the head loss in this system (ft.)?

   Ans. 11.1 ft.

7. A flow of 4 cfs enters, then exits a pipe segment composed of two pipes laid in parallel. The left side of the parallel system is composed of a 6 inch diameter pipe. The right side is composed of a 4 inch diameter pipe which enlarges to an 8 inch diameter pipe. The water velocity in the 8 inch

diameter pipe is 2 ft./sec. What is the water velocity in the 6 inch diameter pipe?

Ans. 16.8 ft/sec.

8. A 6 inch diameter main (C=110) is carrying a flow of 1 cfs. What head loss is produced by two regular flanged 45 degree elbows and a globe valve?

   Ans. 4 ft.

9. Water entering a house through a 1 inch diameter horizontal copper pipe makes two 90 degree turns before the pipe ends at a faucet. If the pressure at entrance is 35 psi and the pipe is 25 ft. long, when the faucet is fully open, what is the flow?

   Ans. .095 cfs

10. A 24 inch diameter transmission main carries 7 cfs for a length of two miles. Along the pipe are three 45 degree elbows and 4 open gate valves. What percentage of the total head loss is caused by the fittings on this pipe?

    Ans. 1.4%

chapter 6

1. Find the Hydraulic Radius of an 18 inch diameter sewer pipe:
   a. flowing full
   b. flowing half full

   Ans. a. .375 ft.
   b. .375 ft.

2. Water flows 6.25 ft. deep in a rectangular canal 8 ft. wide at a velocity of 14.9 ft./sec. What is the slope on which the canal is laid if n=.01?

   Ans. .0031 ft./ft.

3. Determine the drop over 1000 ft. of rectangular wooden flume which carries 600 cfs at a 7 ft. depth. The flume measures 15 ft. across.

   Ans. .4 ft.

4. A rectangular channel whose width is three times its depth is lined with smooth concrete. If flow is 900 cfs and velocity is 3 ft./sec, what is the channel incline?

   Ans. .00009 ft/ft.

5. A 12 inch diameter lined cast iron sewer flowing full is laid on a grade of .004 ft./ft. Calculate the flow.

   Ans. 2.3 cfs

6. A 48 inch diameter circular steel drain is laid on a slope of .003, and is flowing full. What is the flow?

   Ans. 63.9 cfs

7. The cross sectional measurements of a flume are 10 inches by 10 inches. The water is 8 inches deep, and moving at a velocity of 2 ft./sec. How many gallons of water will the flume deliver in 4 hours?

   Ans. 118,483 gal.

8. Water flows three feet deep in a rectangular channel 20 ft. wide ($n=.013$, $s=.0144$). How deep would the same quantity flow on a slope of .00144?

   Ans. 6.5 ft.

9. An obstruction in a grass lined horizontal channel 6 ft. wide causes a depth buildup of 2 ft. behind it. If channel flow is 10 MGD, how deep is the water 1000 ft. upstream?

   Ans. 26 ft.

10. A clay drainpipe flowing full slopes downward 3 ft. over a length of 1500 ft. If the water velocity in the pipe is 5 ft./sec., what is the diameter of the pipe?

Ans. 4 ft.

chapter 7

1. What size orifice is required to discharge .565 cfs under a head of 28.5 ft?

   Ans. 2 inch diam.

2. A venturi meter reads 60 psi above the meter and 40 psi at the throat. Pipe diameter is 8 inches. Throat diameter is 2 inches. Calculate the flow (cfs).

   Ans. 1.17 cfs

3. A 90 degree V-notch weir is to be installed in a 24 inch diameter sewer to measure 500 gpm. What head should be expected?

   Ans. .72 ft.

4. A weir 3 ft. high extends 20 ft. across a rectangular channel in which there is 100 cfs flowing. What is the water depth just upstream from the weir?

   Ans. 4.3 ft.

5. What is the gage difference on a mercury manometer, in inches, for a 3 cfs discharge through a 10 inch by 4 inch venturi meter?

   Ans. 17.5 inches

6. A pitot tube in a pipe in which water is flowing is connected to a differential manometer containing water. If the difference in water levels in the manometer is 3.5 inches, what is the velocity of flow in the pipe?

   Ans. 4 ft./sec.

7. To prevent freezeup, a metered customer is allowed a let run at a rate of 2 gpm. If the customer is not to be billed for this water use, and the utility charges $.03/cubic foot registered on the meter for water, how much should be deducted from the customer's monthly bill while the let run continues?

   Ans. $346.52

8. A utility with 125 miles of water mains produces a flow of 12.12 MGD. There are 8.71 MGD of metered customer sales, .37 MGD of metered use at public buildings, and 141 hours of firefighting per year at 2000 gpm. An average meter is underregistering by 2%. What is the % lost water (unaccounted for)?

   Ans. 26.2%

9. An ice cube factory moves into the city. It needs a 15 gpm flow for the ice cube machine and 25 gpm for the water-cooled compressors. There will also be a rest room and several hose connections. A 2 inch diameter line servicecs the building. A second ice cube machine may be installed in a few years. What meter should be supplied to meet present and future water needs?

   Ans. 2 inch pd or 2 inch cmpd.

10. A single story apartment building will be constructed to house 20 families. It will be on level terrain, at a distance of 300 ft. from an 8 inch diameter supply main with a pressure of 80 psi. Size the service line to the apartment, and install the meter.

    Ans. 2 inch DIP
    2 inch pd meter

chapter 8

1. Referring to the pump curve given:
   a. When the pump is operating under 90 ft. of head, what is the horsepower use?

b. What is the head at which it will run most efficiently?

Ans. a. 9 hp

b. 100 ft.

2. Referring to the pump curve given, what is the delivered flow under each of these conditions:
   a. discharge lift 100 ft.
   discharge piping head loss 15 ft.
   suction head 25 ft.
   suction head losses 10 ft.
   b. discharge lift 50 ft.
   head loss in discharge piping 12 ft.
   suction lift 20 ft.
   suction head loss 13 ft.

   Ans. a. 240 gpm

   b. 0

3. Referring to the pump curve given, what is the motor horsepower needed to run the pump if the motor is 80% efficient and the pump is delivering 140 gpm?

   Ans. 7.5 mhp

4. A pump draws 60 Kw. Head loss is 9 ft.; velocity is 5 ft./sec. Pipe is 3 inch diameter. If wire to water efficiency is 60%, how high can the pump lift water in this system?

   Ans. 1719 ft.

5. Referring to pump curve given, if suction lift is 12 ft. and discharge is through 4000 ft of horizontal line (slope=.015), what would be the flow?

   Ans. 325 gpm

6. If a tank 150 ft. long, 10 ft. deep and 35 ft wide is being dewatered with a pump rated at 100 gpm, and it takes exactly three days to empty the tank, how efficient is the pump?

   Ans. 91%

7. Given two identical pumps, each with the characteristics shown in the pump curve given, what is the discharge for a head of 80 ft:
   a. if both pumps are operated in series
   b. if both pumps are operated in parallel

   Ans. a. 380 gpm
   b. 620 gpm

8. A pump has an impeller diameter of 14 inches and operates at 1750 RPM. What is the velocity of the water circulating at the impeller rim?

   Ans. 107 ft/sec

9. NPSHR for a particular pump is 6 ft. Installed into the system described below, determine the NPSHA, and the probability of efficient operation of the pump.
   Suction lift 15 ft.
   Suction head losses 12 ft.
   Flow 300 gpm
   Suction side inlet 4 inches
   Vapor pressure 2 ft.

   Ans. 6 ft., cavitation likely

10. An industrial process pump is taking 400 gpm from a tank through a 6 inch diameter line from a submerged bottom inlet. The water depth over the inlet is 15 ft; the distance from the tank to the pump is 35 ft.; the pump elevation is 8 ft. above inlet. What is the pressure reading at the suction inlet of the pump?

    Ans. 2.7 psi

Values of C for Hazen-Williams Formula

Type of Pipe	C
Asbestos Cement	140
Brass	140
Brick Sewer	100
Cast Iron	
New, unlined	130
New, lined	140
10 years old	110
20 years old	90
Ductile Iron, (cement lined)	140
Concrete or Concrete Lined	
Smooth, steel forms	140
Wooden forms	120
Rough	110
Copper	140
Fire Hose (rubber lined)	135
Galvanized Iron	120
Glass	140
Lead	130
Masonry Conduit	130
Plastic	150
Steel	
Coal-tar enamel lined	150
New Unlined	140
Riveted	110
Tin	130
Vitrified	120
Wood Stave	120

Hazen-Williams Alignment Chart

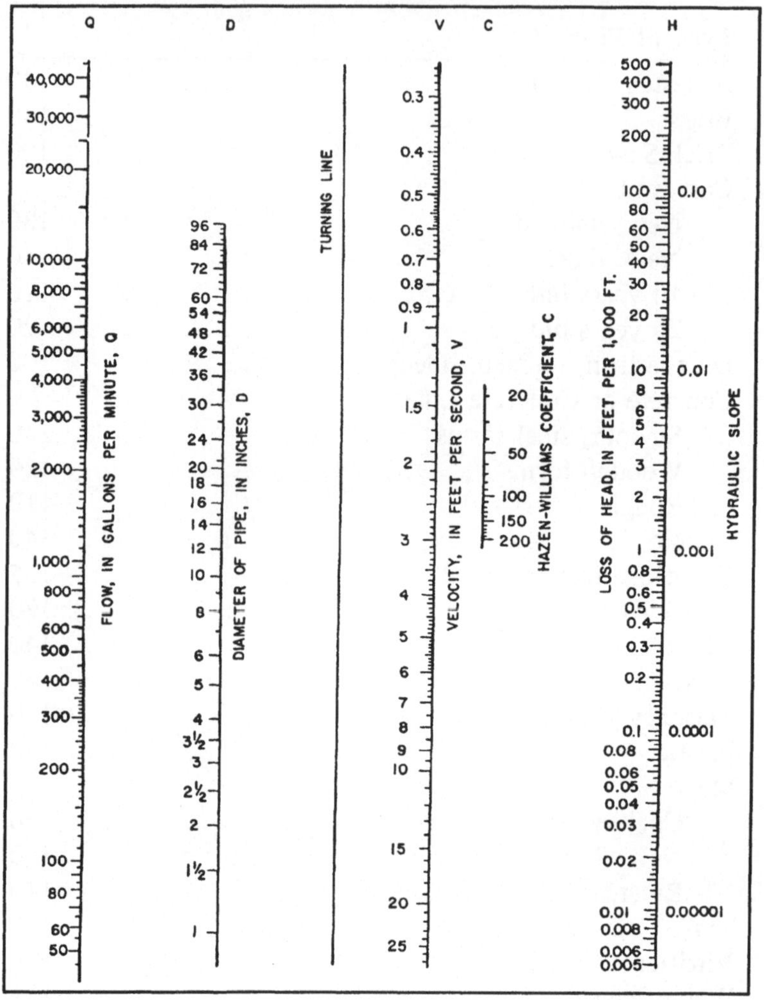

Directions For Using Hazen-Williams Alignment Chart

For determining head loss, find the proper discharge and line it up with the pipe diameter. Extend the line across to the Pivot Line, and mark the point where the Pivot Line was crossed. Line up that point with the proper C factor on the other side, and extend it to the head loss.

Velocity is determined from a line which passes through flow and diameter. Pivot Line is not needed for this. Pivot line is needed only for finding C factor or head loss.

Resistance to Flow from Valves and Fittings

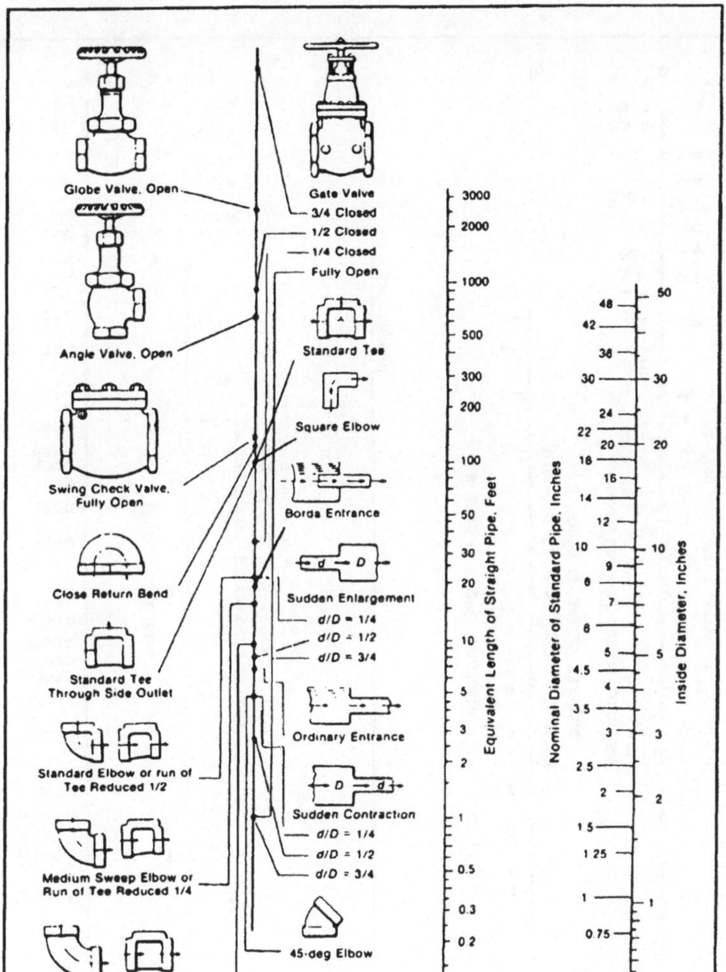

Reprinted with permission of Crane Co.

Manning's Nomograph

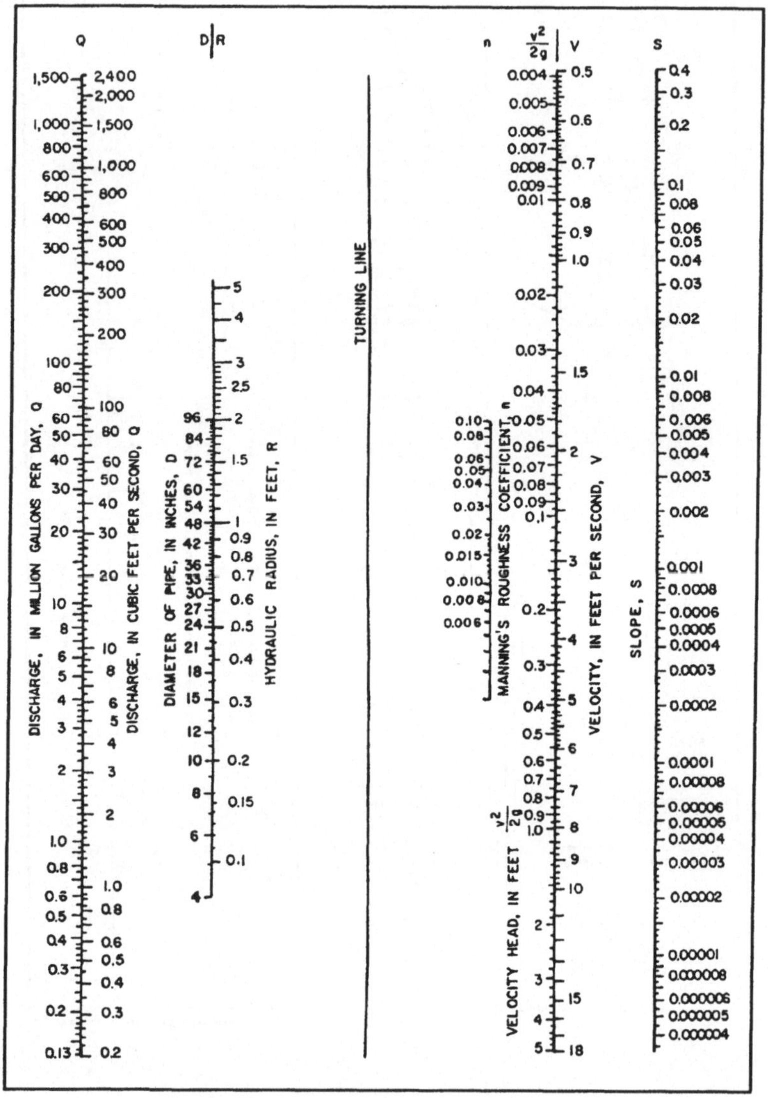

Manning Roughness Coefficient, n

Type of Conduit	n
Pipe:	
Cast Iron, coated	0.012 - 0.014
Cast Iron, uncoated	0.013 - 0.015
Wrought Iron, galvanized	0.015 - 0.017
Wrought Iron, black	0.012 - 0.015
Steel, riveted and spiral	0.015 - 0.017
Corrugated	0.021 - 0.026
Wood Stave	0.012 - 0.013
Cement Surface	0.010 - 0.013
Concrete	0.012 - 0.017
Vitrified	0.013 - 0.015
Clay, drainage tile	0.012 - 0.014
Lined Channels:	
Metal, smooth semicircular	0.011 - 0.015
Metal, corrugated	0.023 - 0.025
Wood, planed	0.010 - 0.015
Wood, unplaned	0.011 - 0.015
Cement Lined	0.010 - 0.013
Concrete	0.014 - 0.016
Cement Rubble	0.017 - 0.030
Grass	----- - 0.200
Unlined Channels:	
Earth: straight and uniform	0.017 - 0.025
Earth: dredged	0.025 - 0.033
Earth: winding	0.023 - 0.030
Earth: stony	0.025 - 0.040
Rock: smooth and uniform	0.025 - 0.035
Rock: jagged and irregular	0.035 - 0.045

Plumbing Fixture Value

Fixture Type	**Fixture Value Based on 35 psi at Meter Outlet**
Bathtub	8
Bedpan washer	10
Combination sink and tray	3
Dental unit	1
Dental lavatory	2
Drinking fountain (cooler)	1
Drinking fountain (public)	2
Kitchen sink: ½-in. connection	3
3/4-in. connection	7
Lavatory: 3/8-in. connection	2
½-in. connection	4
Laundry tray: ½-in. connection	3
3/4-in. connection	7
Shower head (shower only)	4
Service sink: ½-in. connection	3
3/4-in. connection	7
Urinal: Pedestal flush valve	35
Wall or stall	12
Trough (2-ft. unit)	2
Wash sink (each set of faucets)	4
Water closet: Flush valve	35
Tank type	3
Dishwasher: ½-in. connection	4
3/4-in. connection	10
Washing machine: ½-in. connection	5
3/4-in. connection	12
1-in. connection	25
Hose connections (wash down): ½-in.	6
3/4-in.	10
Hose (50-ft. length - wash down): ½-in.	6
5/8-in.	9
3/4-in.	12

Reprinted from Sizing Water Service Lines and Meters, M22, by permission.

Water Flow Demand per Fixture Value - Low Range

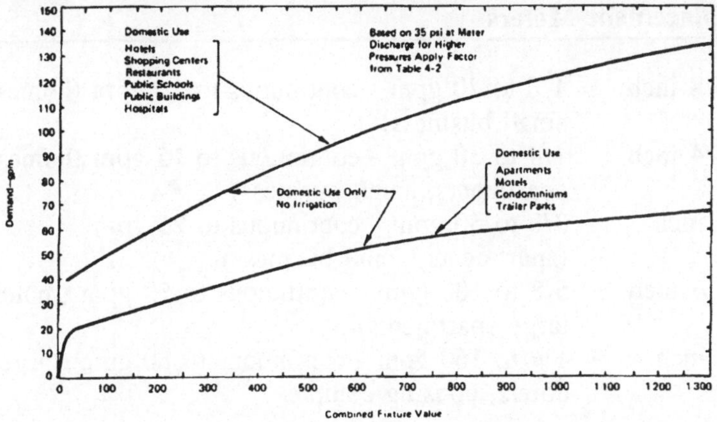

Water Flow Demand per Fixture Value - High Range

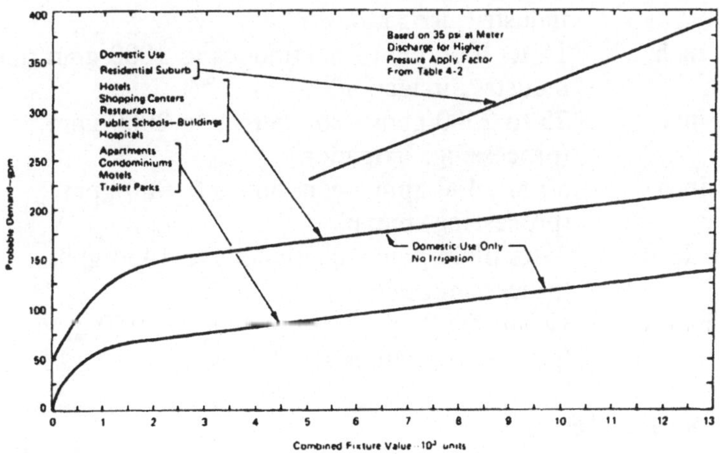

Reprinted from Sizing Water Service Lines and Meters, M22, by permission.
Copyright 1975, American Water Works Association.

Meters By Flow

Displacement Meters

5/8 inch	1/8 to 20 gpm - continuous to 10 gpm (homes, small business)
3/4 inch	1/4 to 30 gpm - continuous to 15 gpm (homes, apartments, small business)
1 inch	3/8 to 50 gpm - continuous to 25 gpm (apartments, small business)
1½ inch	5/8 to 100 gpm - continuous to 50 gpm (motels, large apartments)
2 inch	1¼ to 160 gpm - continuous to 80 gpm (large hotels, housing complex)

Turbine Meters

2 inch	3 to 200 gpm - continuous to 160 gpm (hotels, housing complexes)
3 inch	4 to 450 gpm - continuous to 350 gpm (hotels, industry, irrigation)
4 inch	15 to 1000 gpm - continuous to 1000 gpm (large industry, irrigation)
6 inch	25 to 2500 gpm - continuous to 2000 gpm (processing, irrigation)
8 inch	50 to 2400 gpm - continuous to 900 gpm (processing, pump)
10 inch	75 to 3800 gpm - continuous to 1450 gpm (processing, pumps)
12 inch	120 to 5000 gpm - continuous to 2150 gpm (processing, pumps)

Compound Meters

2 inch	1/4 to 160 gpm - continuous to 160 gpm (motels, schools)
3 inch	1/2 to 350 gpm - continuous to 350 gpm (hotels, large hospitals)

Meters By Flow (continued)

Compound Meters (continued)

4 inch	3/4 to 1000 gpm - continuous to 1000 gpm (apartments, dormitories)
6 inch	1½ to 1000 gpm - continuous to 500 gpm (apartment complexes)
8 inch	2 to 1600 gpm - continuous to 800 gpm (large condominiums)
10 inch	4 to 2300 gpm - continuous to 1150 gpm (large complexes)

Reprinted from Water Audits and Leak Detection, M36, by permission.

Meter Head Loss

Displacement Meters

Size	Pressure Loss (psi at gpm)
5/8 inch	8.0 at 20
3/4 inch	10.0 at 30
1 inch	11.0 at 50
1½ inch	11.5 at 100
2 inch	12.0 at 160

Compound Meters

Size	Pressure Loss (psi at gpm)
2 inch	5.0 at 160
3 inch	4.0 at 320
4 inch	3.0 at 500
6 inch	14.0 at 1000
8 inch	14.5 at 1600

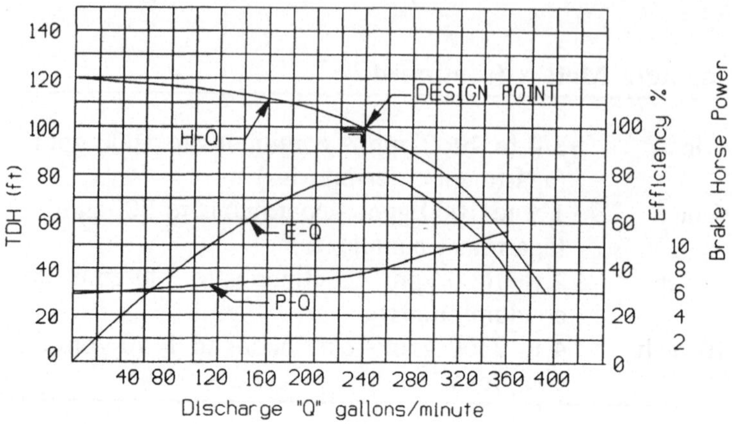

PUMP CHARACTERISTIC CURVE

glossary

Absolute Pressure - the entire pressure on a unit area, including atmospheric pressure; recorded as psia.

Atmospheric Pressure - the pressure of the atmosphere upon the earth; (14.7 psi).

Axial Pump - water enters and exits the pump on the same plane.

Bernoulli's Theorem - neglecting friction losses, from point to point in a fluid system employing steady state flow, the total energy is the same at every point in the path of flow.

Bourdon Tube - a pressure gage; pressure source is connected to a spring assembly which registers on a scale.

Brake Horsepower - the power applied to the pump.

C Factor - a numerical designation of interior pipe roughness.

Cavitation - constant formation and implosion of vapor cavities in a moving liquid, causing damage to pipes, pumps and fittings.

Centrifugal Pump - a pumping mechanism whose rapidly spinning impeller imparts a high velocity to the water which enters, then converts that velocity to pressure upon exit.

Compound Meter - a meter which includes both a positive displacement meter to measure small flows, and a turbine meter to measure large flows.

Compound Pipe Systems - an arrangement of two or more different pipes, either connected end to end, or side by side.

Density - amount of mass per unit volume of a substance.

Double Suction Pump - water entering approaches the impeller from both sides.

Drawdown - the distance the water level drops once pumping begins.

Dynamic Discharge Head - the vertical distance between the pump centerline and the water level on the discharge side, plus the friction head loss.

Dynamic Suction Head - the vertical distance between the water level on the suction side and the centerline of the pump below it, minus the friction head loss.

Dynamic Suction Lift - the vertical distance between the water level and the centerline of the pump above it, plus the friction head loss.

Efficiency - the power produced by the pump, divided by the power used in operating the pump.

Elevation Head - pressure due to the elevation of the water.

Equation of Continuity - ($Q=AV$); at a given flow, water velocity is dependent upon the cross sectional area of flow.

Flow - the quantity of fluid which passes a point in a given unit of time. Flow can be viewed as a moving volume.

Flow Rate Controller - venturi-valve arrangement which automatically controls flow through a constant rate sand filter.

Fluid - matter which will deform without external stress, and takes the shape of its container. Both liquids and gases are fluids. A fluid is infinitely flexible; it can move rapidly in one place and slowly in another. It can transmit a force in any or all directions.

Force - pressure upon an entire area; registered in pounds.

Gage Pressure - the pressure above or below atmospheric pressure; recorded as psig.

Head - pressure, registered in feet.

Head Loss - pressure loss due to friction, registered in feet.

Horizontal Pump - pump shaft is in a horizontal position.

Hydraulic Grade Line - the line that connects the piezometric surfaces along a pipeline.

Hydraulic Radius - wetted area divided by wetted perimeter; a measure of the efficiency with which the conduit can transmit flow.

Hydraulics - the study of the principles that govern the behavior of liquids at rest and in motion. It is the study of the mechanics of water and its control by man.

Laminar Flow - this is ideal fluid flow; water particles move along straight, parallel paths, in layers or streamlines. There is no turbulence in the water, and no friction loss. This is not typical of normal pipe flow because the water velocity is too great, but it does occur in groundwater flow, and an attempt at laminar flow is demonstrated in the use of tube and plate settlers in settling basins.

Major Head Loss - pressure loss due to friction along the length of pipe.

Manometer - a piezometer tube curved into a U shape and filled with an internal fluid of higher specific gravity than the fluid being measured. One or both ends may be connected to a source of pressure.

Mass - the quantity of matter that a substance contains. Matter is divided into two major classes: Solids and Fluids.

Minor Head Loss - pressure loss due to turbulence at bends, valves, fittings and changes in pipe diameter.

Motor Horsepower - the power applied to the motor.

Multi-Stage Pump - more than one impeller on the pump shaft.

Net Positive Suction Head - the total amount of energy, in Absolute feet of water, at the pump centerline.

Open Channel Flow - flow in a conduit which has one side in contact with atmosphere. Open channel flow is caused by the downward slope of the channel, which maintains the velocity. Flow in aqueducts, streams, rivers and sewer pipes is open channel flow.

Orifice Meter - flat plate across the line of flow which passes through a small hole in its center; a pressure differential meter.

Parallel Pipe Systems - two or more pipes of different age, material, size or length, laid side by side, with the flow splitting among them.

Pascal's Law - the pressure at any one point in a static liquid is exerted with equal intensity in all directions.

Parshall Flume - constricted section of an open channel through which the flow passes; reading is by depth measurement.

Piezometer - an open-ended vertical tube attached at the point of pressure measurement; this is the simplest form of pressure gage.

Pitot Gage - a flow measurement device consisting of a double pressure gage, one having an end pointed upstream. Velocity Head is calculated by measurement of pressure differential.

Positive Displacement Meter - measuring chamber encloses a disc or piston which oscillates to pass the flow and transmits the movement to a register, which records.

Power - the rate at which work is done; (ft.lb./sec).

Pressure - an amount of force acting on a unit area.

Pressure Head - pressure due to the depth of water.

Propeller Meter - for measuring mainline flow, chamber encloses a small propeller which responds to velocity.

Proportional Meter - bypasses a small percentage of the flow through the meter and measures it; the measurement is then multiplied up to the actual flow.

Pump Characteristics - the range of heads, flows, horsepowers, and efficiencies at which a particular pump will operate.

Pump Performance Curve - a diagram of how a particular pump will perform over the range of its characteristics.

Pump System - the arrangement of pipe, fittings, and equipment through which the pumped water will flow.

Pumping Water Level - water level when the pump is on.

Radial Pump - flow leaves the pump at a 90 degree angle from the entrance direction.

Radius of Influence - horizontal distance between pump shaft and outermost area of drawdown.

Safe Well Yield - maximum pumping rate achievable without increasing the drawdown.

Series Pipe Systems - two or more pipes of different age, material or size, laid end to end.

Single Stage Pump - one impeller on the pump shaft.

Single Suction Pump - one suction line enters the pump casing.

Slope - in closed pipe systems, the feet of head loss for each foot of pipe.

Solid - matter which holds its shape and will not deform without external stress.

Specific Capacity - pumping rate per foot of drawdown; (gpm/ft.)

Specific Gravity - density of a substance compared to the density of a standard substance.

Static Discharge Head - the vertical distance from the pump centerline to the water level on the discharge side.

Static Suction Head - the vertical distance from the water surface on the suction side to the pump centerline below it.

Static Suction Lift - the vertical distance from the water surface on the suction side to the pump centerline above it.

Static Water Level - water level when the pump is off.

Steady State Flow - flow is steady if at any one point the velocity of the water does not vary and the pressure is constant.

Submersible Pump - vertical, close-coupled centrifugal pump designed to work with both motor and pump submerged in the water being pumped.

Surge - slow motion mass tendency of the water to "pile up" as momentum decreases; a pressure increase results.

System Curve - a curve representing the head that the system will create over a range of pumped flows.

Total Dynamic Head - the difference between the dynamic suction head and the dynamic discharge head, or the sum of the dynamic suction lift and the dynamic discharge head.

Total Static Head - the difference between the static suction head and the static discharge head, or the sum of the static suction lift and the static discharge head.

Thrust - transient pressure created at turns in a pipeline; tends to push the fitting loose.

Turbine Meter - flow passes through a chamber enclosing a rotor which turns in response to velocity.

Turbulent Flow - turbulent flow is normal for a water system; water particles move in a haphazard fashion and continually cross each other in all directions; pressure losses result along a length of pipe because of it.

Vapor Pressure - the water's tendency to evaporate.

Variable Speed Pump - shaft rotation speed varies with system needs.

Velocity - the speed at which a liquid moves.

Velocity Head - pressure energy due to the velocity of the water.

Venturi Meter - a section of constricted pipe through which the flow passes; a pressure differential meter.

Vertical Pump - pump shaft is in a vertical position.

Water Hammer - temporary increase in pressure resulting from rapid change in velocity or direction of the water.

Water Horsepower - the power necessary to lift water to a required height.

Weir - flat plate across the line of flow in an open channel, and over which the flow passes; reading is by depth measurement.

Wire To Water Efficiency - pump efficiency multiplied by motor efficiency.

bibliography

American Water Works Association. *Basic Science Concepts and Applications*. Denver: American Water Works Association, 1980.

American Water Works Association. *Hydraulics and Water Loss Control*. Denver: American Water Works Association, 1984.

American Water Works Association. *Sizing Water Service Lines and Meters*. Denver: American Water Works Association, 1975.

American Water Works Association. *Water Meters: Selection, Installation, Testing, Maintenance*. Denver: American Water Works Association, 1972.

Arasmith, E. *Applied Practical Hydraulics*. Albany: Linn-Benton Community College, 1978.

Brown, E. *Hydraulics For Operators*. Stoneham: Butterworth, 1985.

French, R. *Open Channel Hydraulics*. New York: McGraw-Hill, 1985.

Karassic, I. *Centrifugal Pump Clinic*. New York: Marcel Dekker, 1981.

Kanen, J. *Applied Hydraulics*. New York: Holt, Rinehart & Winston, 1986.

Lambeck, R. *Hydraulic Pumps and Motors*. New York: Marcel Dekker, 1983.

Parr, B. *Treatment Plant Hydraulics For Environmental Engineers*. Englewood Cliffs: Prentice-Hall, 1984.

Vesilind, P., and Pierce, J. *Environmental Engineering*. Boston: Butterworth, 1982.

Walker, R. *Pump Selection*. Ann Arbor: Ann Arbor Science, 1972.

Water Pollution Control Federation. *Design of Wastewater and Stormwater Pumping Stations*. Washington DC: Water Pollution Control Federation, 1981.

index

absolute pressure 14, 15, 102, 103, 135
atmospheric pressure 4, 14, 15, 55, 101-103, 135, 138
axial pump 90, 135

bearings 92, 102, 111
Borda entrance 51, 54
Bourdon tube 12, 135
brake horsepower 38, 94, 99, 118, 135
bubbler 42

cavitation 101-103, 105, 106, 108, 125, 135
characteristic curve 93, 95, 99
coefficient of discharge 65, 66
coefficient of friction 44
compound meters 77, 84, 85, 133, 134
compound pipes 49
cone of depression 41

datum 27
density 1-3, 7, 110, 111, 137, 142
design point 95-97
discharge head 34-36, 116, 137, 142, 144
double suction 90, 137
drawdown 40-42, 103, 118, 137, 142
dynamic discharge head 35, 36, 116, 137, 144
dynamic suction head 35, 36, 137, 144
dynamic suction lift 36, 137, 144
dynamic water system 21, 33

elevation head 22-24, 26-28, 43, 114, 137
entrained air 105, 106, 109
equation of continuity 5, 6, 137

fire flow 73, 78
flow rate controller 67, 138

gage position 13
groundwater 9, 20, 40, 41, 138

Hazen-Williams alignment chart 48, 127

horizontal pump 90, 138
hydraulic grade line 19-21, 30, 35, 43, 56, 119, 138
hydraulic jump 57
hydraulic radius 58, 120, 138

impeller eye 88, 89, 103

jet pumps 108

laminar flow 4, 138, 139
lift pumps 33, 101
lift station 107

magnetic flow meter 74
Manning's formula 58, 60
manometers 68, 69
mechanical seals 91
motor horsepower 38, 118, 124, 140

net positive suction head 102, 103, 140
nomograph 49, 51, 53, 60, 61, 129

ordinary entrance 51, 54
ordinary exit 51, 54
orifice meters 64, 66, 70

packing 91, 105, 111
Parshall flumes 71, 72
Pascal's Law 9, 14, 140
peak flow demand 79
performance curve 92, 93, 103, 141
piezometer 12, 19, 20, 31, 68, 72, 140, 141
piezometric surface 19-21
piston pump 87
pitot 25, 73, 122, 141
positive displacement meters 75, 76
pressure differential meters 64, 70
pressure head 21-28, 43, 56, 65, 73, 88, 114, 141
priming 100, 101
propeller meters 78
proportional meters 78
pump curve 94, 109, 123-125
pump heads 33
pumping test 42
pumping water level 40, 41, 141

radius of influence 41, 142
reducer 105, 106
rotameter 75
roughness coefficient 47, 58, 130

safe well yield 40, 142
scouring velocity 5, 56, 60
screw pump 87
service lines 78, 82, 83, 131, 132, 143
shutoff head 95, 109
specific capacity 41, 118, 142
specific gravity 2, 3, 68, 69, 109, 111, 140, 142
specific weight 1, 7
standpipe 10, 23, 113
static discharge head 34, 35, 142, 144
static suction head 34, 142, 144
static suction lift 34, 35, 142, 144
static water level 40, 41, 143
static water system 20, 33

steady state flow 4, 5, 26, 55, 60, 135, 144
stilling well 72, 75
stuffing box 91, 111
submersible pumps 89, 106-108
suction head 34-36, 100, 102, 103, 106, 109, 124, 125, 137, 140, 142, 144
suction lift 34-36, 51, 100, 102, 103, 108, 109, 116, 124, 125, 137, 142, 144
surge 17, 95, 144
system curve 96, 97, 144

thrust 17, 18, 92, 144
total dynamic head 36, 94, 116, 118, 144
total static head 34, 35, 144
totalizer 64, 75
trajectory 72
turbine meters 76, 84, 133

ultra sonic meters 74
ultrasonic detectors 42

vacuum 14, 36, 40, 68, 96, 100, 113
vacuum pumps 100
vapor pressure 101-104, 109, 125, 144
variable speed pumps 99, 100
velocity head 23-29, 44, 47, 53-55, 65, 73, 88, 103, 108, 114, 141, 145
velocity head technique 53
venturi 28, 29, 66-68, 71, 108, 122, 138, 145
vertical pump 89, 145

water hammer 16, 17, 145
water horsepower 38, 116, 118, 145
weirs 70, 71
well pumps 89, 98, 108
wetted area 58, 138
wetted perimeter 58, 138
wire to water efficiency 39, 118, 124, 145

notes

notes

notes

notes

notes

notes

notes

notes

notes

notes

notes

notes

notes